algebra for all orange level

elizabeth warren PhD

ORIGO
EDUCATION

About the Author

Elizabeth Warren has been involved in Mathematics Education for more than 30 years. During this time she has actively worked in both the university and school levels and engaged both elementary and secondary schoolteachers in professional development activities. She is presently conducting research into Patterns and Algebra.

Algebra for All, *Orange Level*

Copyright 2006 ORIGO Education
Author: Elizabeth Warren, PhD

Warren, Elizabeth.
Algebra for All: orange level
ISBN 1 921023 02 3.
1. Algebra - Problems, exercises, etc. - Juvenile literature. I. Title.
512

For more information, contact
North America
Tel. 1-888-ORIGO-01 or 1-888-674-4601
Fax 1-888-674-4604
sales@origomath.com
www.origomath.com

Australasia
For more information,
email info@origo.com.au
or visit www.origo.com.au for other contact details.

ISBN: 1 921023 02 3

10 9 8 7 6 5 4 3 2 1

INTRODUCTION

What is algebra?

Algebraic thinking commences as soon as students identify consistent change and begin to make generalizations. Their first generalizations relate to real-world experiences. For example, a child may notice a relationship between her age and the age of her older brother. In the example below, Ali has noted that her brother Brent is always 2 years older than her.

Ali's age	Brent's age
8	10
9	11
10	12
11	13

Over time these generalizations extend to more abstract situations involving symbolic notation that includes numbers. The above relationship can be generalized using the following symbolic notation.

$$\textbf{Ali + 2 = Brent} \qquad \textbf{A + 2 = B}$$

Algebraic thinking uses different symbolic representations, such as unknowns and variables, with numbers to explore, model, and solve problems that relate to change and describe generalizations. The symbol system used to describe generalizations is formally known as algebra. Following the example above, Ali wonders how old she will be when Brent is 21 years old. We can solve a problem such as this by "backtracking" the generalization ($A = 21 - 2$).

Algebra involves the generalizations that are made regarding the relationships between variables in the symbol system of mathematics.

Why algebra?

Identifying patterns and making generalizations are fundamental to all mathematics, so it is essential that students engage in activities involving algebra. Many practical uses for algebra lie hidden under the surface of an increasingly electronic world, such as specific rules used to determine telephone charges, track bank accounts and generate statements, describe data represented in graphs, and encrypt messages to make the Internet secure. Algebraic thinking is more overt when we create rules for spreadsheets or simply use addition to solve a subtraction problem.

What are the "big ideas"?

The lessons in the *Algebra for All* series aim to develop the "big ideas" of early algebra while supporting thinking, reasoning, and working mathematically. These ideas of equivalence and equations, patterns and functions, properties, and representations are inherent in all modern curricula and are summarized in the following paragraphs.

Equivalence and Equations

The most important ideas about equivalence and equations that students need to understand are:

- "Equals" indicates equivalent sets rather than a place to write an answer
- Simple real-world problems with unknowns can be represented as equations
- Equations remain true if a consistent change occurs to each side (the balance strategy)
- Unknowns can be found by using the balance strategy.

Patterns and Functions

This idea focuses on mathematics as "change". Change occurs when one or more operation is used. For example, the price of an item bought on the Internet changes when a freight charge is added. It is important for students to understand that:

- Operations almost always change an original number to a new number
- Simple real-world problems with variables can be represented as "change situations"
- "Backtracking" reverses a change and can be used to solve unknowns.

Properties

Students will discover a variety of arithmetic properties as they explore number, such as:

- The commutative law and the associative law exist for addition and multiplication but not for subtraction and division
- Addition and subtraction are inverse operations, as are multiplication and division
- Adding or subtracting zero and multiplying or dividing by 1 leaves the original number unchanged
- In certain circumstances, multiplication and division distribute over addition and subtraction.

Representations

Different representations deepen our understanding of real-world problems and help us identify trends and find solutions. This idea focuses on creating and interpreting a variety of representations to solve real-world problems. The main representations that are developed in this series include graphs, tables of values, drawings, equations, and everyday language.

INTRODUCTION

About the series

Each of the six *Algebra for All* books features 4 chapters that focus separately on the "big ideas" of early algebra — Equivalence and Equations, Patterns and Functions, Properties, and Representations. Each chapter provides a carefully structured sequence of lessons. This sequence extends across the series so that students have the opportunity to develop their understanding of algebra over a number of years.

About the lessons

Each lesson is described over 2 pages. The left-hand page describes the lesson itself, including the aim of the lesson, materials that are required, clear step-by-step instructions, and a reflection. These notes also provide specific questions that teachers can ask students, and subsequent examples of student responses. The right-hand page supplies a reproducible blackline master to accompany the lesson. The answers for all blackline masters can be found on pages 66-73.

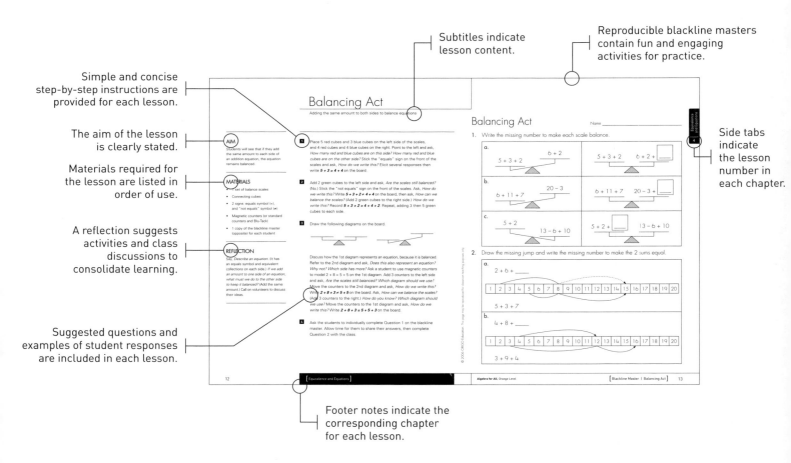

Subtitles indicate lesson content.

Reproducible blackline masters contain fun and engaging activities for practice.

Simple and concise step-by-step instructions are provided for each lesson.

The aim of the lesson is clearly stated.

Materials required for the lesson are listed in order of use.

A reflection suggests activities and class discussions to consolidate learning.

Suggested questions and examples of student responses are included in each lesson.

Side tabs indicate the lesson number in each chapter.

Footer notes indicate the corresponding chapter for each lesson.

Assessment

Students' thinking is often best gauged by the conversations that occur during classroom discussions. Listen to your students and make notes about their thinking. You may decide to use the rubric below to assess students' mathematical proficiency in the tasks for each lesson. Study the criteria, then assess and record each student's understanding on a copy of the Assessment Summary provided on page 74. Although the summary lists every lesson in this book, it is not necessary to assess students for all lessons.

A	The student fully accomplishes the purpose of the task. Full understanding of the central mathematical ideas is demonstrated. The student is able to communicate his/her thinking and reasoning.
B	The student substantially accomplishes the purpose of the task. An essential understanding of the central mathematical ideas is demonstrated. The student is generally able to communicate his/her thinking and reasoning.
C	The student partially accomplishes the purpose of the task. A partial or limited understanding of the central mathematical ideas is demonstrated and/or the student is unable to communicate his/her thinking and reasoning.
D	The student is not able to accomplish the purpose of the task. Little or no understanding of the central mathematical ideas is demonstrated and/or the student's communication of his/her thinking and reasoning is vague or incomplete.

Building Balances

Using balance to represent addition equations

AIM

Students will represent addition equations using balance scales and extend this understanding to different representations.

MATERIALS

- 1 set of balance scales

- Connecting cubes

- Sign showing the equals symbol (=)

- Magnetic counters (or standard counters and Blu-Tack)

- 1 copy of the blackline master (opposite) for each student

REFLECTION

Discuss what makes an equation. Ask, *What does an equation need to have?* (An equals symbol and equivalent collections on each side.) *Is 4 + 6 = 7 + 8 an equation?* (No.) *Why not?* (It does not have equivalent collections on each side.) *Is 4 + 5 + 2 = 6 + 5 an equation?* (Yes.) *Why?* (It has an equals symbol and equivalent collections on each side.)

1 Place 6 red cubes and 2 yellow cubes on one side of the balance scales. Ask, *How many ways can we balance the scales?* Call on volunteers to use cubes and the "equals" sign to model their answers. Each time, invite another student to write the matching equation on the board.

2 Draw the following diagram on the board.

Ask a volunteer to use the magnetic counters to model "4 add 3 equals 5 add 2". Then invite another student to write the matching equation on the board. (4 + 3 = 5 + 2)

3 Introduce a 3rd method of representing an equation. Write **5 + 2** on the left side of the diagram and **4 + 3** on the right as shown.

Ask, *How can we check that this equation is balanced?* (Check that there are equivalent collections on each side.) Invite volunteers to explain their ideas.

4 Allow time for the students to complete the blackline master and share their answers.

[Equivalence and Equations]

Building Balances

Name _____

1. Draw counters to balance the scales and complete the matching sentence or number sentence for each.

Model	Language	Number sentence
a.	Four add four equals six add two.	____ + ____ = ____ + ____
b.	_____ _____ _____	8 + 3 = 6 + 5

2. Write numbers to balance the scales then complete the missing parts.

a. 5 + 6 8 + 3	Five add six equals eight add three.	____ + ____ = ____ + ____
b.	Two add five equals six add one.	____ + ____ = ____ + ____
c.	_____ _____ _____	1 + 8 = 4 + 5

Feathered Friends

Exploring equivalence involving addition and subtraction

AIM

Students will use pictures to model equivalent real-world addition and subtraction stories, and express these stories in symbolic form.

MATERIALS

- Magnetic counters (or standard counters and Blu-Tack)
- Counters for each student
- 1 copy of the blackline master (opposite) for each student

REFLECTION

Give each student a handful of counters. Have them each draw 2 trees on a sheet of blank paper. Then challenge them to write equivalence stories.

1 Draw 2 ponds on the board. Use counters to model the story. Say, *There are 3 ducks in one pond. Another 2 ducks join them. There are 6 ducks in the other pond. One of these ducks flies away. How many ducks are in the first pond?* (3 + 2 = 5) *How many ducks are in the second pond?* (6 – 1 = 5) *Does each pond have the same number of ducks?* (Yes.) Elicit several responses then ask, *How can I show this?* Write **3 + 2 = 6 – 1** on the board.

2 Discuss the word "equation". Explain that equations have equivalent collections on each side of the "equals" symbol. Highlight the similarities between the words "equation" and "equals". Repeat Step 1 for **8 – 3 = 10 – 5** and **7 + 2 + 2 = 13 – 2**. Each time, invite volunteers to tell a story incorporating 2 ponds and ask other students to write each story as an equation.

3 Read the blackline master with the class. Make sure the students understand that for each question, each tree represents one side of the equation. Allow time for the students to complete the questions and share their answers.

Feathered Friends

Name _____

Write numbers to show what you see in each tree picture.

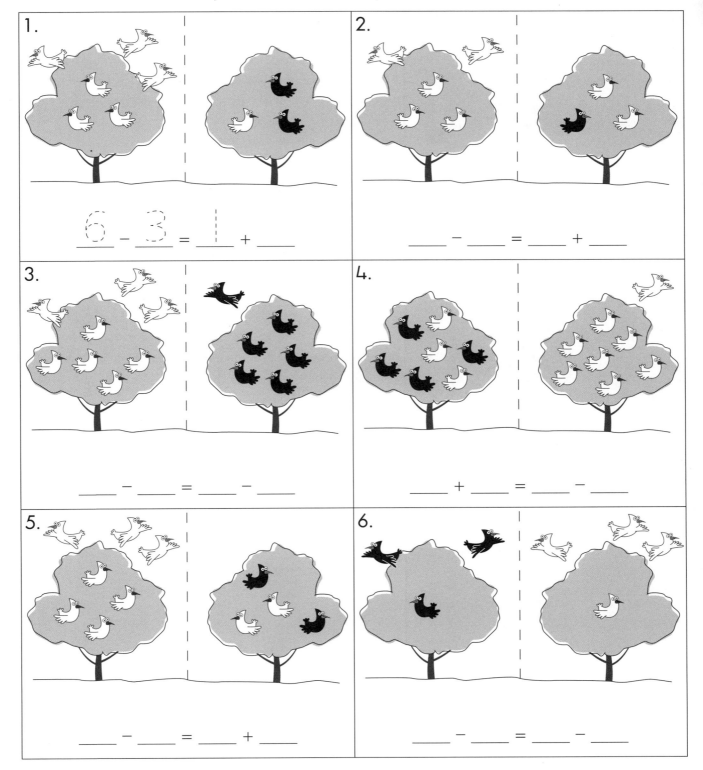

1.

$6 - 3 = 1 +$ ___

2.

___ − ___ = ___ + ___

3.

___ − ___ = ___ − ___

4.

___ + ___ = ___ − ___

5.

___ − ___ = ___ + ___

6.

___ − ___ = ___ − ___

Same Sides

Using balance to represent addition and subtraction equations

AIM

Students will extend the representations for addition equations to include subtraction equations.

MATERIALS

- 1 set of balance scales
- Connecting cubes
- Sign showing the equals symbol (=)
- Blu-Tack
- 1 copy of the blackline master (opposite) for each student

REFLECTION

Ask, *Is 24 – 8 + 12 = 30 – 3 an equation?* (No.) *Why not?* (It does not have equivalent collections on each side.) *Is 45 = 66 – 23 + 2 an equation?* (Yes.) *Why?* (It has an equals symbol and equivalent collections on each side.) Invite several volunteers to share their thinking.

1 Write **7 + 4 = 13 – 2** on the board. Ask a volunteer to use the cubes and the "equals" sign to model "7 add 4 equals 13 subtract 2" on the scales. Discuss how one side of the model represents 7 + 4 = 11, and how it is difficult to see that the other side shows 13 – 2 = 11. Ask, *How can we model this equation more clearly?* Call on volunteers to share their ideas.

2 Draw the following diagram on the board.

Ask, *How can we show the equation using this diagram?* Elicit several responses. Encourage the students to use symbols, as shown.

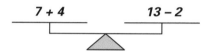

Ask, *How do we know the equation is equal?* (There are equivalent collections on each side.)

3 Repeat Steps 1 and 2 for **5 + 8 = 16 – 3** and **17 – 5 = 3 + 9**.

4 Ask the students to complete the blackline master. Make sure they understand that each question has 3 parts that match. Call on volunteers to share their answers with the class.

[Equivalence and Equations]

Same Sides

Name _____

Write the missing parts.

Model	Language	Number Sentence
1. 12 + 9 10 + 11	12 add 9 equals 10 add 11.	___ + ___ = ___ + ___
2.	33 equals 29 add 4.	___ = ___ + ___
3. 34 − 5 27 + 2	_____ _____ _____	___ − ___ = ___ + ___
4.	67 subtract 9 equals 56 add 2.	___ − ___ = ___ + ___
5.	_____ _____ _____	34 + 15 = 55 − 6

Balancing Act

Adding the same amount to both sides to balance equations

AIM

Students will see that if they add the same amount to each side of an addition equation, the equation remains balanced.

MATERIALS

- 1 set of balance scales
- Connecting cubes
- 2 signs: equals symbol (=), and "not equals" symbol (≠)
- Magnetic counters (or standard counters and Blu-Tack)
- 1 copy of the blackline master (opposite) for each student

REFLECTION

Say, *Describe an equation.* (It has an equals symbol and equivalent collections on each side.) *If we add an amount to one side of an equation, what must we do to the other side to keep it balanced?* (Add the same amount.) Call on volunteers to discuss their ideas.

1 Place 5 red cubes and 3 blue cubes on the left side of the scales, and 4 red cubes and 4 blue cubes on the right. Point to the left and ask, *How many red and blue cubes are on this side? How many red and blue cubes are on the other side?* Stick the "equals" sign on the front of the scales and ask, *How do we write this?* Elicit several responses then write **5 + 3 = 4 + 4** on the board.

2 Add 2 green cubes to the left side and ask, *Are the scales still balanced?* (No.) Stick the "not equals" sign on the front of the scales. Ask, *How do we write this?* Write **5 + 3 + 2 ≠ 4 + 4** on the board, then ask, *How can we balance the scales?* (Add 2 green cubes to the right side.) *How do we write this?* Record **5 + 3 + 2 = 4 + 4 + 2**. Repeat, adding 3 then 5 green cubes to each side.

3 Draw the following diagrams on the board.

Discuss how the 1st diagram represents an equation, because it is balanced. Refer to the 2nd diagram and ask, *Does this also represent an equation? Why not? Which side has more?* Ask a student to use magnetic counters to model 2 + 8 = 5 + 5 on the 1st diagram. Add 3 counters to the left side and ask, *Are the scales still balanced? Which diagram should we use?* Move the counters to the 2nd diagram and ask, *How do we write this?* Write **2 + 8 + 3 ≠ 5 + 5** on the board. Ask, *How can we balance the scales?* (Add 3 counters to the right.) *How do you know? Which diagram should we use?* Move the counters to the 1st diagram and ask, *How do we write this?* Write **2 + 8 + 3 = 5 + 5 + 3** on the board.

4 Ask the students to individually complete Question 1 on the blackline master. Allow time for them to share their answers, then complete Question 2 with the class.

[Equivalence and Equations]

Balancing Act

Name _____

1. Write the missing number to make each scale balance.

<table>
<tr><td>

a.

6 + 2

5 + 3 + 2

</td><td>

5 + 3 + 2   6 + 2 + [___]

</td></tr>
<tr><td>

b.

20 − 3

6 + 11 + 7

</td><td>

6 + 11 + 7   20 − 3 + [___]

</td></tr>
<tr><td>

c.

5 + 2

13 − 6 + 10

</td><td>

5 + 2 + [___]   13 − 6 + 10

</td></tr>
</table>

2. Draw the missing jump and write the missing number to make the 2 sums equal.

a.

2 + 6 + _____

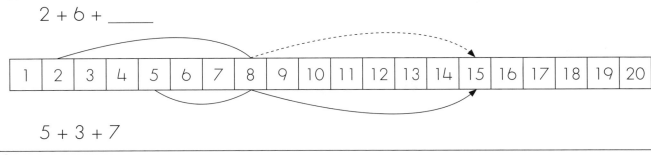

5 + 3 + 7

b.

4 + 8 + _____

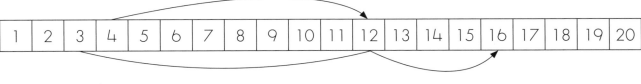

3 + 9 + 4

Keeping Balanced

Subtracting the same amount from both sides to balance equations

AIM

Students will see that if they subtract the same amount from both sides of an addition equation, the equation remains balanced.

MATERIALS

- 1 set of balance scales

- Connecting cubes

- Magnetic counters (or standard counters and Blu-Tack)

- 1 copy of the blackline master (opposite) for each student

REFLECTION

Ask, *If we subtract an amount from one side of an equation, what must we do to the other side to keep the equation balanced?* (Subtract the same amount.) Ask several students to share their thinking.

1 Place 6 red cubes and 2 blue cubes on the left side of the scales, and 5 red cubes and 3 blue cubes on the right. Ask, *How many red and blue cubes are on each side of the scales? How do we write this?* Elicit responses then write **6 + 2 = 5 + 3** on the board. Remove 2 red cubes from the left and ask, *Are the scales still balanced? How do we write this?* Record **6 + 2 − 2 ≠ 5 + 3** on the board. Ask, *How can we balance the scales?* Call on volunteers to share their answers, then remove 2 red cubes from the right. Ask, *How can we write this?* Write **6 + 2 − 2 = 5 + 3 − 2** on the board. Repeat, subtracting 3 then 5 red cubes from each side.

2 Draw the following diagrams on the board.

Ask, *Which diagram represents an equation? Why?* Elicit responses then ask a volunteer to use magnetic counters to model 6 + 5 = 9 + 2 on the 1st diagram. Remove 4 counters from the left side and ask, *Are the scales still balanced? Which diagram should we use?* Move the counters to the 2nd diagram. Ask, *How do we write this?* Write **6 + 5 − 4 ≠ 9 + 2** on the board. Ask, *How can we make this balance?* Remove 4 counters from the right and ask, *Is the equation balanced now? How do you know? Which diagram should we use?* Move the counters to the 1st diagram and ask, *How do we write this?* Elicit responses, then write **6 + 5 − 4 = 9 + 2 − 4** on the board.

3 Ask the students to complete Question 1 on the blackline master. Read Question 2 with the class then allow time for them to complete the question. Call on volunteers to share their answers.

[Equivalence and Equations]

Keeping Balanced

Name _____

1. Write the missing number to make each scale balance.

a.

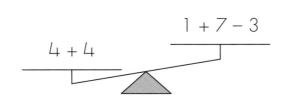

$$4 + 4 \qquad 1 + 7 - 3$$

$$4 + 4 - \boxed{\underline{\qquad}} \qquad 1 + 7 - 3$$

b.

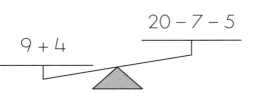

$$9 + 4 \qquad 20 - 7 - 5$$

$$9 + 4 - \boxed{\underline{\qquad}} \qquad 20 - 7 - 5$$

c.

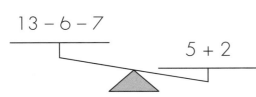

$$13 - 6 - 7 \qquad 5 + 2$$

$$13 - 6 - 7 \qquad 5 + 2 - \boxed{\underline{\qquad}}$$

2. Draw the missing jump and write the missing number to make the 2 sums equal.

a.

$$2 + 6 - \underline{\qquad}$$

| 1 | 2 | 3 | 4 | 5 | 6 | 7 | 8 | 9 | 10 | 11 | 12 | 13 | 14 | 15 | 16 | 17 | 18 | 19 | 20 |

$$5 + 3 - 2$$

b.

$$4 + 11 - \underline{\qquad}$$

| 1 | 2 | 3 | 4 | 5 | 6 | 7 | 8 | 9 | 10 | 11 | 12 | 13 | 14 | 15 | 16 | 17 | 18 | 19 | 20 |

$$3 + 12 - 4$$

Secret Shapes

Investigating conventions for symbols for unknowns

AIM

Students will explore the conventions for representing unknowns.

MATERIALS

- Small boxes in 2 different shapes, such as hearts and cylinders (1 of each for each student)

- 1 handful of small flat beads or counters for each student

- 1 copy of the blackline master (opposite) for each student

REFLECTION

Ask, *If 2 unknowns are the same, what values must they have?* (Values must be the same.) *If there are different unknowns, what values might they have?* (Values can be the same or different.)

1 Show 2 identical heart-shaped boxes and say, *These boxes are the same shape. Can they cost the same amount? Can they cost different amounts?* Elicit several responses then discuss the convention of "same shape, same value" for the unknown in an equation. Write $\heartsuit + \heartsuit = 8$ on the board. Ask, *What is the value of each heart?* (4) *Can they have different values?* (No.) Write $\heartsuit + \heartsuit = 22$ on the board and ask, *What is the value of each heart in this equation?* (11)

2 Show 1 of each shaped box and say, *These boxes are different shapes. Can they cost the same amount?* (Yes.) Discuss the idea that different shapes can have the same or different values. Write $\heartsuit + \bigcirc = 11$ on the board and draw the horizontal table shown below.

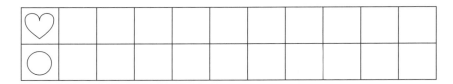

Ask the students to suggest different possible values for the heart and circle. Record their values in the table. Each time, have the students use the beads and corresponding boxes to check their solution. Tell them that each bead represents "1". Ask, *How do we know that we have all the possible answers?* Discuss how to make an ordered list (0, 11; 1, 10; 2, 9; 3, 8; ...). Ask, *Are there any other possible solutions?* (Some students may know that decimal numbers, such as 1.10 and 9.90, can also be solutions.)

3 Direct the students to complete the blackline master. Allow them to use the beads and boxes to help if necessary. Call on volunteeers to share their answers.

Secret Shapes

Name _____

1. Complete the table to show all the possible values for the heart and the circle.

$$\heartsuit + \bigcirc = 8$$

♡									
○									

2. **a.** Complete the table to show all the possible values for the square and the triangle.

$$\square + \triangle = 7$$

□								
△								

b. Check your answers by writing them in this table. The first one is done for you.

□	△	Check
0	7	0 + 7 = 7

Shape Match

Substituting in equations with unknowns

AIM

Students will revisit the conventions for representing unknowns, and substitute values in equations with unknowns on both sides. They will use the balance strategy to find values for the unknowns.

MATERIALS

- 1 copy of the blackline master (opposite) for each student

REFLECTION

Ask, *If the unknowns are the same, what values must they have?* (They each must have the same value.) *If the unknowns are different, what values can they have?* (They can each have the same value or different values.) Call on volunteers to share their ideas.

1 Write $\heartsuit + \square + 3 = \heartsuit + 5$ on the board. Ask, *What value must the hearts have in this equation?* (The hearts must each have the same value.) *If the heart has a value of 2, how can we figure out the value of the square?* Elicit responses, then substitute 2 for each heart in the equation. Write $2 + \square + 3 = 2 + 5$ below the original equation. Then make the calculations and write $\square + 5 = 7$. Ask, *How can we figure out the value for the square?* Encourage the students to make suggestions and work together to solve the equation. Then ask, *What are some different values for the heart?* With each answer, find the value of the square using the balance method. Remind the students that an equation is kept balanced by adding or subtracting the same amount from each side.

2 Draw the following diagram on the board. Say, *The value of the heart is 3 and the value of the square is 2. Is $\square + 5 + \heartsuit = 6 + \square + \square$ a true equation? What are some more true equations?* Call on a volunteer to write an equation on the board. Ask, *How can we check that the equation is true?* (Substitute 3 for the heart and 2 for the square.) Invite several students to write equations. Each time, have another student make the substitutions to check the equation.

| $\heartsuit$ | 3 |
| $\square$ | 2 |

3 Read the blackline master with the class. Allow time for the students to complete the questions and share their answers.

[Equivalence and Equations]

Shape Match

1. Find possible solutions for this equation.

$$\heartsuit + \heartsuit + \square = 12$$

a.	b.	c.
If $\heartsuit = 2$	If $\heartsuit = 3$	If $\heartsuit = 0$
$2 + 2 + \square = 12$	___ + ___ + $\square$ = 12	___ + ___ + $\square$ = 12
Then $\square$ = ____	Then $\square$ = ____	Then $\square$ = ____

2. Write 8 different equations using these values. $\heartsuit = 8$ $\square = 2$

Then write number sentences to check your equations.

Equation	Check
$\heartsuit + \square + 2 = 12$	$8 + 2 + 2 = 12$

More or Less

Using "more than", "greater than", and "less than"

AIM

Students will see that if **a** is greater than **b**, then **b** is less than **a**.

MATERIALS

- 1 set of balance scales
- Connecting cubes
- 3 signs: "greater than", "more than", and "less than"
- Blu-Tack
- 1 copy of the blackline master (opposite) for each student

REFLECTION

Ask, *If 12 is greater than 3, is 3 greater than 12 or less than 12? If ♡ is greater than ◯, is ◯ greater than ♡ or less than ♡?* Call on volunteers to share and justify their reasoning.

1 Place 10 red cubes on the left side of the scales and 8 blue cubes on the right. Ask, *Which side has more? How can we write this?* Call on volunteers to share their ideas then write **10 is more than 8** on the board. Stick the "more than" sign on the front of the scales. Ask, *How else can we say this?* Call on volunteers to share their solutions, then write **10 is greater than 8** below **10 is more than 8**. Stick the "greater than" sign on the front of the scales. Turn the scales around so that the 8 cubes are on the left side. Ask, *How can we write this?* Invite several responses then write **8 is less than 10** below **10 is greater than 8**. Stick the "less than" sign on the front of the scales. Repeat for **12** and **6**.

2 Place 6 red cubes and 3 blue cubes on the left side of the scales, and 3 red cubes and 4 blue cubes on the right. Ask, *How many cubes do we have on the left?* (6 add 3, or 9.) *How many cubes on the right?* (3 add 4, or 7.) *Which side has more? How do we write this?* Write **6 add 3 is more than 3 add 4** on the board. Ask, *What is another way we can write this?* Call on volunteers to share their ideas then write **6 add 3 is greater than 3 add 4** on the board. Turn the scales around and ask, *How can we write this?* Encourage several responses then write **3 add 4 is less than 6 add 3** on the board.

3 Have the students complete the blackline master. Call on volunteers to share their responses.

[Equivalence and Equations]

More or Less

Name _____

1. Complete the sentences to match the numbers on the scales.

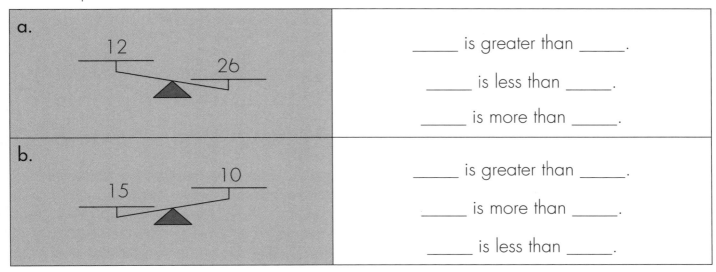

a.

12   26

_____ is greater than _____.

_____ is less than _____.

_____ is more than _____.

b.

15   10

_____ is greater than _____.

_____ is more than _____.

_____ is less than _____.

2. For each of these, draw counters on the scales. Then complete the sentence.

a.

6 + 4 is greater than 2 + 3.

_____ + _____ is less than _____ + _____.

b.

9 + 3 is less than 12 + 2.

_____ + _____ is greater than _____ + _____.

3. Write some numbers on the scales. Then complete the sentences to match.

a.

_____ is greater than _____.

_____ is less than _____.

b.

_____ is less than _____.

_____ is greater than _____.

Pattern of Threes

Completing patterns in multiples of three

AIM

Students will relate repeating patterns to patterns of three.

MATERIALS

- Numeral cards for 1 to 20
- Red and blue counters
- 1 copy of the blackline master (opposite) for each student

REFLECTION

Ask, *Will 52 be in this pattern of threes?* (No.) *How do you know?* (Its digits add to 7.) *What is the largest number that you can think of that will be in this pattern of threes?* Call on volunteers to share their ideas.

1 Seat the students on the floor and distribute the numeral cards for 1 to 20. Have the students with the cards for 1 to 10 place them in a line, in order, to make a number track. Then ask them to use the red and blue counters to make a "red, red, blue" pattern below the number track.

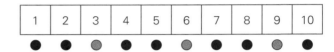

Say, *Let's read aloud the numbers that are above the blue counters.* (3, 6, 9) *What could we call this pattern?* (A pattern of threes.) Discuss why this is a pattern of threes.

2 Ask the students with the cards for 11 to 20 to extend the number track to 20. Ask, *What other numbers are in this pattern of threes? How do you know? Will 16 have a red or blue counter? Will 19 have a red or blue counter?* Call on volunteers to use the counters to complete the pattern. Then have the students read aloud the pattern of threes. (3, 6, 9, 12, 15, 18)

3 Ask the students to complete the blackline master. Call on volunteers to share their answers. Examine the numbers in the pattern of threes and ask, *How can you quickly decide if a number is in the pattern of threes?* (Check if the sum of its digits is either 3, 6, or 9.)

Pattern of Threes

Shade every third number.

| Start | 1 | 2 | 3 | 4 | 5 | 6 |

| 13 | 12 | 11 | 10 | 9 | 8 | 7 |

| 14 | 15 | 16 | 17 | 18 | 19 | 20 |

| 27 | 26 | 25 | 24 | 23 | 22 | 21 |

| 28 | 29 | 30 | 31 | 32 | 33 | 34 |

| 40 | 39 | 38 | 37 | 36 | 35 |

Patterns and Functions

1

Repeating Repeating

Relating repeating patterns to repeated addition

AIM

Students will explore how repeating patterns are repeated addition, and use this to find the shapes of blocks in different positions.

MATERIALS

- Pattern blocks
- 5 copies of the numeral card for 3
- 1 copy of the blackline master (opposite) for each student

REFLECTION

Ask, *If a pattern has 6 blocks in each repeat, and the 6th block is blue, what other blocks in the pattern must be blue?* (12th, 18th, ...) *If the 5th block is yellow, what other blocks must be yellow?* (10th, 15th, ...) Ensure the students justify their answers.

1 Seat the students on the floor. Have them use the pattern blocks to make the repeating pattern shown below.

Call on a volunteer to separate the pattern into its repeating parts. Ask, *How many blocks are in each repeat?* (3) Below each repeat, place a numeral card for 3. Point to the 1st shape (square) and say, *This is the 1st shape.* Point to the 3rd shape (triangle) and ask, *What position is this shape?* (3rd) *Which shape comes 9th in the pattern?* (Triangle.) *How do you know?* (Counting from the 1st shape, or adding three 3s.) *How can we figure out what the 14th shape will be?* Discuss different ways, for example, by counting from the 1st shape, 5 more than the 9th shape, or adding four 3s and then adding 2.

2 Ask the students to complete the blackline master. Discuss how the students figured out the answers. For Question 1, some students may count by 4s and then count on, for example, for the 14th shape — 4, 8, 12, then count on 2 shapes into the next repeat. Some may count by 4s and then count back, for example, for the 14th shape — 4, 8, 12, 16, then count back 2.

Repeating Repeating

Name _____

1. a. Write the number of shapes in each repeat. The first one is done for you.

___4___   _____   _____   _____   _____

    b. Draw the 12th shape. _____   **c.** Draw the 20th shape. _____

    d. Draw the 14th shape. _____   **e.** Draw the 17th shape. _____

    f. Imagine the pattern continues. Draw the 45th shape. _____

    Write how you figured it out. _____

    _____

    _____

2. a. Write the number of shapes in each repeat.

_____   _____   _____

    b. Draw the 9th shape. _____   **c.** Draw the 5th shape. _____

    d. Draw the 15th shape. _____   **e.** Draw the 11th shape. _____

    f. Imagine the pattern continues. Draw the 31st shape. _____

    Write how you figured it out. _____

    _____

    _____

Changing Faster

Relating repeating patterns to rates of change

AIM

Students will record simple repeating patterns in a table and describe which component of the pattern is growing at a faster rate.

MATERIALS

- Pattern blocks
- 1 sheet of paper
- 1 copy of the blackline master (opposite) for each student

REFLECTION

Say, *Describe a repeating pattern where the number of each shape increases at the same rate.* (A pattern with the same number of each shape in each repeat.) *Describe a repeating pattern where the number of each shape increases at different rates.* (A pattern with a different number of each shape in each repeat.) Call on several volunteers to share their ideas.

1 Seat the students on the floor close to the board. Ask a student to use the pattern blocks to make the pattern shown below.

Ask, *Is this a growing pattern or a repeating pattern? How do you know? What is the repeating part?* Invite students to extend the pattern for 4 more repeats.

2 Ask, *What shapes are in the pattern?* Draw a table on the board, as shown on the right.

▲	□

Cover part of the pattern with a sheet of paper so that the students can only see the first repeat. Ask, *How many triangles and how many squares can you see?* (1 triangle and 3 squares.) Record this in the first row of the table. Continue until the whole pattern is revealed. Each time, record the number of each shape, as shown on the right.

▲	□
1	3
2	6
3	8
4	12

Then refer to the table and ask, *How many more triangles and squares are there each time?* (1 triangle and 3 squares.) *Which is increasing faster, the number of triangles or the number of squares?* (The number of squares.)

3 Complete Question 1 on the blackline master as a class. Then allow time for the students to complete Question 2 individually. Invite volunteers to share and explain their answers.

[Patterns and Functions]

Changing Faster

1. a. Loop the repeating part of this pattern.

b. Write the number of triangles and squares you see in each row below.

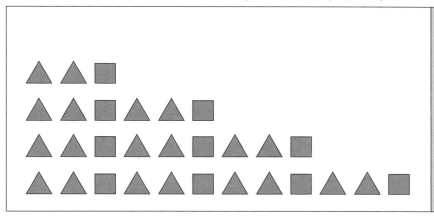

	▲	■
Row 1	2	1
Row 2		
Row 3		
Row 4		

c. There are _____ more triangles and _____ more square each time.

d. Draw the shape that is increasing faster. _____

2. a. Loop the repeating part of this pattern.

b. Complete this table.

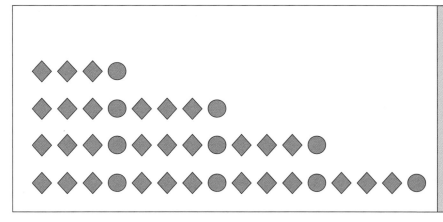

	◆	●
Row 1	3	1
Row 2		
Row 3		
Row 4		

c. There are _____ more diamonds and _____ more circle each time.

d. Draw the shape that is increasing faster. _____

Growing Bigger

Predicting patterns according to position

AIM

Students will relate growing patterns to position patterns, and use position to predict the pattern.

MATERIALS

- Identical magnetic counters (or standard counters and Blu-Tack)

- 1 copy of the blackline master (opposite) for each student

REFLECTION

Ask, *How do you know if a pattern is growing or repeating?* Call on volunteers to share their thinking.

1 Use counters to make the following pattern on the board and label it as shown below.

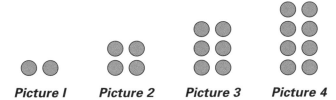

Picture 1 **Picture 2** **Picture 3** **Picture 4**

Ask, *Is this a growing pattern or a repeating pattern? How do you know?* Invite several responses.

2 Point to the width and height of the 1st picture and say, *Describe Picture 1.* (2 counters wide and 1 counter high.) *Describe Picture 2* (2 wide and 2 high), *Picture 3* (2 wide and 3 high), *and Picture 4* (2 wide and 4 high). *Imagine the next picture in the pattern. How wide will it be?* (2) *How high will it be?* (5) Discuss how the width is always 2 and the height is the same as the picture number. Ask, *What will Picture 6 look like? What will Picture 10 look like?* Call on volunteers to extend the pattern to Picture 10 to check the answer.

3 Refer to Question 1a on the blackline master. Call on volunteers to describe each picture. Then say, *Describe Picture 5.* (5 wide and 1 high.) *Use the picture number to help you describe the pattern?* (The width of each picture is the same as the picture number.) Allow time for the students to complete the question. Call on volunteers to share their answers. Repeat the discussion for Questions 2 and 3 before asking the students to complete the blackline master.

Growing Bigger

Name _____

1. a. Draw Picture 5 and Picture 10 in this pattern.

Picture 1   Picture 2   Picture 3   Picture 4   Picture 5   Picture 10

b. Use the picture number to help you describe the pattern.

_____

_____

2. a. Draw Picture 5 and Picture 10 in this pattern.

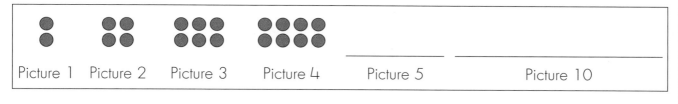

Picture 1   Picture 2   Picture 3   Picture 4   Picture 5   Picture 10

b. Describe the pattern.

_____

_____

3. a. Draw Picture 5 and Picture 10 in this pattern.

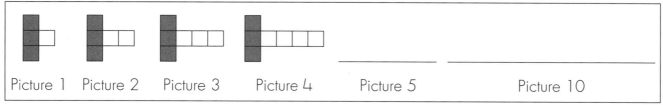

Picture 1   Picture 2   Picture 3   Picture 4   Picture 5   Picture 10

b. Describe the pattern.

_____

_____

Simple Steps

Predicting steps in a growing pattern

AIM

Students will identify growing patterns and predict steps in the pattern design.

MATERIALS

- Red and blue tiles

- 7 signs: "Picture 1", "Picture 2", "Picture 3", "Picture 4", "Picture 5", "Picture 6", and "Picture 10"

- 1 copy of the blackline master (opposite) for each student

REFLECTION

Refer to the pattern in Question 1 on the blackline master. Ask, *How can we quickly figure out how many red tiles are in a picture? Is it helpful to look at the number of tiles in Picture 1? Do we need to look at more than one picture?* Invite volunteers to share their ideas.

1 Seat the students on the floor and use the tiles to make the pattern shown below.

Picture 1   **Picture 2**   **Picture 3**   **Picture 4**

Place the matching sign below each picture in the pattern and ask, *How many red tiles are in Picture 1? How many red tiles are in Picture 4?* Invite several responses. Place the signs for Pictures 5 and 6 at the end of the pattern, then ask, *How many red tiles will be in Picture 5? How many in Picture 6?* Invite a student to make the red component of these 2 pictures. Ask, *Are there more red tiles than the picture number?* (No.) *Is the number of red tiles the same as the picture number?* (Yes.) *How many red tiles will be in Picture 10?* Place the "Picture 10" sign on the floor and ask a student to make the red component. Repeat the discussion for the blue components.

2 Ask, *Which picture will have 8 red tiles and 16 blue tiles?* (Picture 8.) *Which picture will have 15 red tiles and 30 blue tiles?* (Picture 15.) Ask volunteers to extend the pattern to check the answers.

3 Direct the students to work in pairs to complete the blackline master. Allow time for them to share and justify their answers. Afterward, for each of the two questions, ask the students which picture will have 12 ☐ and 13 ■.

Simple Steps

1.

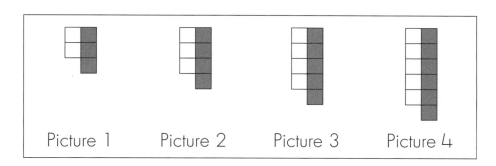

Picture 1   Picture 2   Picture 3   Picture 4

a. How many ☐ are in Picture 1? _____ Picture 2? _____ Picture 3? _____

How many ■ are in Picture 1? _____ Picture 2? _____ Picture 3? _____

Imagine the pattern continues.

b. How many ☐ will be in Picture 5? _____ Picture 10? _____

How many ■ will be in Picture 5? _____ Picture 10? _____

2.

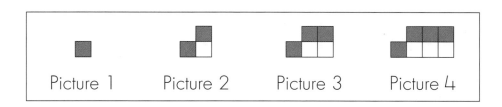

Picture 1   Picture 2   Picture 3   Picture 4

a. How many ☐ are in Picture 1? _____ Picture 2? _____ Picture 3? _____

How many ■ are in Picture 1? _____ Picture 2? _____ Picture 3? _____

Imagine the pattern continues.

b. How many ☐ will be in Picture 5? _____ Picture 6? _____ Picture 10? _____

How many ■ will be in Picture 5? _____ Picture 6? _____ Picture 10? _____

Number Patterns

Increasing and decreasing number patterns

AIM

Students will identify and extend number patterns involving both addition and subtraction sequences.

MATERIALS

- 1 copy of the blackline master (opposite) for each student

- Calculator for each student

REFLECTION

Ask, *How can we tell if a number pattern is growing or repeating? How do we figure out the increase in an addition number pattern? How do we figure out the decrease in a subtraction number pattern?* Call on volunteers to share their solutions.

1 Write the addition number pattern **5, 15, 25, 35, 45, 55** on the board. Ask, *What is the 1st number in this pattern? What is the 2nd number in this pattern? Is the pattern growing or repeating?* Invite several responses then ask, *What do you do to the 1st number to make the 2nd number?* (Add 10.) *What do you do to the 2nd number to make the 3rd number? By how much is each number growing?* Record the answers on the board as shown.

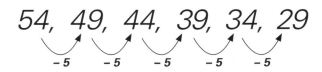

Call on volunteers to write the next 4 numbers in the pattern.

2 Write the subtraction number pattern **54, 49, 44, 39, 34, 29** on the board. Ask, *What is the 1st number in this pattern? What is the 2nd number?* Invite several responses then ask, *What do you do to the 1st number to make the 2nd number?* (Subtract 5.) *What do you do to the 2nd number to make the 3rd number? By how much is each number decreasing?* Record the answers on the board as shown below.

$$54, \ 49, \ 44, \ 39, \ 34, \ 29$$
$$-5 \quad -5 \quad -5 \quad -5 \quad -5$$

Call on volunteers to write the next 4 numbers in the pattern.

3 Read the blackline master with the class. Have the students work in pairs to complete the questions. Allow them to use calculators to check their answers.

Number Patterns

Name _____

1. Write the amount that is being added, then complete the pattern.

a.

4, 8, 12, 16, 20, _____, _____, _____, _____, _____

b.

8, 11, 14, 17, 20, _____, _____, _____, _____, _____

c. Which pattern is growing faster? _____

2. Complete the subtraction pattern.

a.

123, 119, 115, 111, 107, _____, _____, _____, _____, _____

b.

99, 89, 79, 69, 59, _____, _____, _____, _____, _____

c. Which pattern is decreasing faster? _____

3. **a.** Write an addition pattern.

23, _____, _____, _____, _____, _____

b. Write a subtraction pattern.

141, _____, _____, _____, _____, _____

Gift Wrap

Using functions to develop an addition rule

AIM

Students will calculate the output number when given the input number and the addition rule.

MATERIALS

- 1 medium-sized open box
- Sign showing the addition symbol (+)
- 2 sets of numeral cards for 1 to 30
- Blu-Tack
- 1 copy of the blackline master (opposite) for each student

REFLECTION

Ask, *What does a function machine do? What does the change rule tell us?* Have the students draw a simple function machine. Tell them the cost of wrapping is $2.50. Have the students write three IN numbers for their machine. Tell them to swap machines with another student and each write the matching OUT numbers.

1 Stick the "addition" sign and the number symbol card for 3 on the front of the box. Say, *This is a function machine. It changes numbers according to the rule on the front. We call this the "change rule".* Call on a volunteer to act as the function machine. Give that student one set of the numeral cards. Distribute some of the remaining cards to several other students. Ask those students to place their cards (IN numbers), one at a time, inside the "machine". Direct the "function machine" to give back the correct OUT card for each IN card. Repeat for several examples and with different students acting as the function machine. Ask, *What does the machine do? What does the "change rule" tell us?* Call on several volunteers to share their ideas.

2 Read the blackline master with the class. Refer to Question 1 and ask, *What is the change rule?* (+ $3) *Imagine I buy a gift for $17 and have it wrapped. How much will I spend in total?* ($20) Have the students complete the blackline master. Allow time for them to share their answers.

Gift Wrap

Name _____

1. Gift wrapping costs $3.
   Write the price with wrapping.

Gift price	Price with wrapping
$63	
$125	
$252	

2. Gift wrapping costs $5.
   Write the price with wrapping.

Gift price	Price with wrapping
$57	
$131	
$324	

3. Read the rule. Write the OUT number.

a.
10 IN +20 OUT ____

b.
21 IN +11 OUT ____

c.
48 IN +12 OUT ____

d.
93 IN +4 OUT ____

e.
17 IN +30 OUT ____

f.
23 IN +13 OUT ____

g.
124 IN +6 OUT ____

h.
30 IN +100 OUT ____

Gifts Galore

Using functions to develop backtracking for the addition rule

AIM

Students will calculate the input number when given the output number and the addition rule.

MATERIALS

- 1 medium-sized open box
- Sign showing the addition symbol (+)
- 2 sets of numeral cards for 1 to 30
- Blu-Tack
- 1 copy of the blackline master (opposite) for each student

REFLECTION

Ask, *What is the change rule? How do we reverse the change rule?* (Use the inverse operation.) Have the students draw a simple function machine. Tell them the cost of wrapping is $4.25. Have the students write three OUT numbers for their machine. Tell them to swap machines with another student and each write the matching IN numbers. Ask, *What is the rule for finding OUT numbers when given IN numbers?* (Use the change rule.) *What is the rule for finding IN numbers when given OUT numbers?* (Reverse the change rule.)

1 Stick the "addition" sign and the numeral card for 3 on the front of the box. Say, *This is a function machine. What is the change rule?* (+ 3) Ask a student to act as the function machine. Give that student one set of the numeral cards. Distribute some of the remaining cards to several other students. Ask those students to place their cards (IN numbers), one at a time, inside the "machine". Each time, direct the "function machine" to give back the correct OUT card.

2 Select 5 of the remaining numeral cards. Say, *Now these are the OUT numbers. How can we figure out the IN numbers? Look at the change rule. Do we add 3?* Invite several responses. Ensure the students understand that they need to reverse the change rule and subtract 3 to figure out the IN numbers. Call on volunteers to calculate the IN numbers.

3 Read the blackline master with the class. Refer to Question 1 and ask, *What is the change rule?* (+ $5) Say, *Imagine I buy a gift and have it wrapped. If I spend a total of $12, how much was the gift?* ($7) *How do you know?* Ask the students to complete the blackline master. Allow time for them to share their IN numbers with the class.

Gifts Galore

1. Gift wrapping costs $5.
Write the gift price.

Gift price	Price with wrapping
	$67
	$131
	$324

2. Gift wrapping costs $3.
Write the gift price.

Gift price	Price with wrapping
	$45
	$230
	$72

3. Read the rule. Write the IN number.

a.
____ IN +20 OUT 50

b.
____ IN +11 OUT 30

c.
____ IN +15 OUT 110

d.
____ IN +50 OUT 135

e.
____ IN +7 OUT 116

f.
____ IN +10 OUT 106

g.
____ IN +12 OUT 111

h.
____ IN +4 OUT 102

Patterns and Functions

8

More Gifts

Finding the rule of a function

AIM

Students will figure out the rule, when given the input and output numbers.

MATERIALS

- 1 copy of the blackline master (opposite) for each student

REFLECTION

Have the students draw a function machine and a table showing five IN and OUT numbers for their machine. Tell them to make sure the rule is not shown. Then have each of the students swap machines and figure out their partner's rule.

1 Say, *Imagine I buy a gift for $7 and have it wrapped. If I spend $12 in total, how much did it cost to wrap the gift?* ($5) Repeat for $25 and $30, $76 and $71, and $63 and $58, at the same store. Ask, *How much does gift wrapping cost?* Read Question 1 on the blackline master with the class. Allow time for the students to complete the question. Call on volunteers to share their answers.

2 Draw a function machine and a table on the board, as shown below.

IN	10	21	12	16	40
OUT	19	30	21	25	49

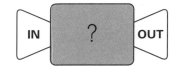

Ask, *How much do we add to the IN number to get the OUT number?* (9) *How much do we subtract from the OUT number to get the IN number?* (9) *What is the change rule?* (+ 9) *How do you know?* Elicit several responses. Write **+ 9** on the front of the function machine. Repeat for sets of IN and OUT numbers to match a "+ 30" rule and then a "+ 12" rule.

3 Allow time for the students to complete the blackline master and share their answers with the class.

More Gifts

Name _____

1. For each of these, figure out the cost of gift wrapping.

a.

Cost of wrapping _____

Gift price	Price with wrapping
$47	$67
$111	$131
$304	$324

b.

Cost of wrapping _____

Gift price	Price with wrapping
$25	$32
$30	$37
$76	$83

2. Write the rules.

a.

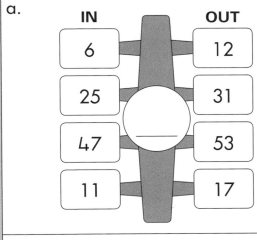

IN	OUT
6	12
25	31
47	53
11	17

b.

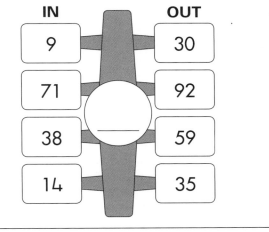

IN	OUT
9	30
71	92
38	59
14	35

c.

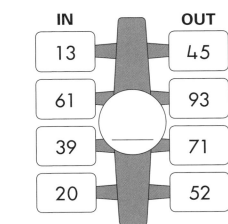

IN	OUT
13	45
61	93
39	71
20	52

d.

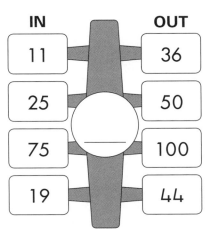

IN	OUT
11	36
25	50
75	100
19	44

Super Savings

Using functions with a subtraction rule

AIM

Students will calculate the output number when given the input number and the subtraction rule.

MATERIALS

- 1 medium-sized open box
- Sign showing the subtraction symbol (–)
- 2 sets of numeral cards for 1 to 30
- Blu-Tack
- 1 copy of the blackline master (opposite) for each student

REFLECTION

Challenge the students to draw function machines for different discounts and list the IN numbers. Ask them to swap machines with another student and each write the matching OUT numbers.

1 Stick the "subtraction" sign and the numeral card for 4 on the front of the box. Say, *This is a function machine. What is the change rule?* Invite a volunteer to act as the function machine. Give that student one set of the numeral cards. Distribute some of the remaining cards to several other students. Ask those students to place their IN cards, one at a time, inside the "machine". Each time, direct the "function machine" to show the correct OUT card. Repeat for several examples and with different students acting as the function machine.

2 Say, *Imagine that the price of a meal is $12 and we have a coupon for $3 off. How much will we pay for the meal?* ($9) Make sure the students know that a coupon will give them a certain amount off the starting price. Repeat for $32, $58, and $41.

3 Ask the students to complete the blackline master. Invite several students to share their answers.

Super Savings

1. A restaurant gives out $4 coupons. Write the new price of each meal.

Meal price	New price
$23	
$57	
$41	

2. A restaurant gives out $7 coupons. Write the new price of each meal.

Meal price	New price
$25	
$39	
$56	

Patterns and Functions

10

3. Read the rule. Write the OUT number.

a. 40 IN −20 OUT ____

b. 61 IN −11 OUT ____

c. 27 IN −5 OUT ____

d. 55 IN −30 OUT ____

e. 117 IN −10 OUT ____

f. 98 IN −8 OUT ____

g. 48 IN −12 OUT ____

h. 30 IN −15 OUT ____

Meal Deals

Using functions to develop backtracking for the subtraction rule

AIM

Students will calculate the input number when given the output number and the subtraction rule.

MATERIALS

- 1 medium-sized open box
- Sign showing the subtraction symbol (–)
- 1 set of numeral cards for 1 to 30
- Blu-Tack
- 1 copy of the blackline master (opposite) for each student

REFLECTION

Challenge the students to draw a function machine for a discount of $2.50 and write the OUT numbers. Have them swap their machines and each write the matching IN numbers. Challenge the students to write a rule for finding IN numbers when given OUT numbers.

1 Stick the "subtraction" sign and the numeral card for 4 on the front of the box. Say, *This is a function machine. What is the change rule?* Distribute some of the numeral cards to several students. Say, *These are the OUT numbers. How can we figure out the IN numbers?* (Add 4.) *How do you know?* Call on several volunteers to share their thinking. Ask those students with numeral cards to place the OUT cards, one at a time, inside the machine. Each time, direct a different student to calculate the correct IN number. Repeat until every student has had a turn to calculate the IN number.

2 Read the blackline master with the class. Make sure the students understand that the full meal price is the price before the coupon is used. Refer to Question 1 and ask, *What is the change rule?* Call on volunteers to share their answers. Say, *Imagine that you used a coupon and paid $25 for a meal. What was the meal price before the $5 was taken off?* ($30) *How do you know?* (Add $5.) Repeat for $15, $35, and $47. Allow time for the students to complete the blackline master. Call on students to share and justify their answers.

Meal Deals

1. A coupon gives $5 off meals.
   Write the full price of each meal.

Full meal price	Price with coupon
	$67
	$31
	$24

2. A coupon gives $3 off meals.
   Write the full price of each meal.

Full meal price	Price with coupon
	$45
	$30
	$72

3. Read the rule. Write the IN number.

a.

_____ IN -10 OUT 47

b.

_____ IN -21 OUT 96

c.

_____ IN -4 OUT 107

d.

_____ IN -12 OUT 89

e.

_____ IN -30 OUT 75

f.

_____ IN -9 OUT 96

g.

_____ IN -6 OUT 195

h.

_____ IN -17 OUT 88

Zero Identity

Subtracting a number from itself to give zero

AIM

Students will conclude that when a number is subtracted from itself the answer is zero.

MATERIALS

- Magnetic counters (or standard counters and Blu-Tack)

- Large number track for 0 to 20

- 1 copy of the blackline master (opposite) for each student

- Calculator for each student

REFLECTION

Say, *Describe some everyday examples of subtracting a number or an amount from itself.* Some examples include emptying a bathtub or sink, spending all the money in our wallet or purse, eating all our lunch. Say, *Describe what happens when the answer to a subtraction sentence is zero.*

1 Say, *Imagine that we have $12 and then we spend $12 on a book. How much money do we have left? If we borrow 5 books from the library and then return them all, how many library books do we still have?* Call on several volunteers to share their solutions.

2 Place 8 counters on the board. Ask, *How many counters are there?* (8) *If we subtract 8 counters* (remove 8 counters), *how many are left?* Encourage the students to describe this action as "8 – 8 = 0". Repeat for 10 and 12. Ask the students to describe what is happening. (When you subtract a number from itself the answer is zero.) Direct the students to complete Question 1 on the blackline master. Call on volunteers to share their answers.

3 Place the number track on the floor. Call on a volunteer to walk from 0 to 14, taking one step in each numbered space. Ask, *How many steps did (Emma) take?* (14) *Let's subtract 14 from that number.* Have the student model this by walking from 14 back to 0 then ask, *What is the answer? How do we write this?* Write **14 – 14 = 0** on the board. Repeat for 11 and 19, using different students each time. Ask the students to complete Question 2.

4 Ask a volunteer to select a number less than 1000. Have the students enter this number into their calculators. Say, *If you subtract the same number, what will happen?* Repeat for 2 or 3 more large numbers. Direct the students to complete the blackline master. Invite volunteers to share their responses.

[Properties]

Zero Identity

Name _____

1.

a. Cross out 17 counters. Write the matching number sentence.

_____ – _____ = 0

b. Cross out $8. Write the matching number sentence.

$1 $1 $1 $1 $1 $1 $1 $1

_____ – _____ = _____

c. Draw 10 balls. Cross out 10 balls. Write the matching number sentence.

_____ – _____ = _____

2.

a. Draw jumps on the number line to show 23 – 23 = 0.

```
<--|||||||||||||||||||||||||||||||||||||||||-->
   0    5   10   15   20   25   30   35   40
```

b. Draw jumps to show 37 – 37 = 0.

```
<--|||||||||||||||||||||||||||||||||||||||||-->
   0    5   10   15   20   25   30   35   40
```

3. Check each of these on your calculator and shade true or false.

a. 56 – 56 = 0	True	False	**b.** 38 – 28 = 0	True	False

4. Write numbers to make these true. Check them on your calculator.

a. _____ – _____ = 0   **b.** _____ – _____ = 0

Piggy Bank

Adding and subtracting the same number to leave the sum unchanged

AIM

Students will determine that when a number is added to another number, and then the same number is subtracted, the answer is the starting number.

MATERIALS

- Counters for each student

- Magnetic counters (or standard counters and Blu-Tack)

- Large number track for 0 to 20

- 1 copy of the blackline master (opposite) for each student

- Calculator for each student

REFLECTION

Ask, *If we add and subtract the same amount from a number, why does that number remain unchanged?* (Subtracting a number from itself gives 0.) *What is a real-world situation where this might occur?* (For example, money in a piggy bank.)

1 Draw the outline of a piggy bank on the board. Invite a student to model this story using the magnetic counters. Say, *Imagine we have $9 and we save $6 more. How much money do we have? How do we write this?* Write **$9 + $6 = $15** on the board. Ask, *If we then spend $6, what will happen?* Invite volunteers to share their ideas. Then ask, *How much money will we have? How do we write this?* Elicit several responses then write **$9 + $6 – $6 = $9** on the board. Repeat for **$10 + $4 – $4** and **$12 + $5 – $5**. Each time ask the students to explain what is happening.

2 Place the number track on the floor. Call on a volunteer to walk from 0 to 14, taking one step in each numbered space. Ask, *How many steps did (Ezra) take? Let's add 3.* Instruct the student to model the addition by walking from 14 to 17. Ask, *What number do we now have? Let's subtract 3.* Instruct the student to model the subtraction then ask, *What number do we have now? How do we write this?* Write **14 + 3 – 3 = 14** on the board. Repeat for **11 + 6 – 6** and **8 + 7 – 7**.

3 Ask the students to complete Questions 1 and 2 on the blackline master. Call on volunteers to share their answers.

4 Ask for a volunteer to select a number less than 1000. Have the students enter this number into their calculator. Ask, *If you add 23 and then subtract 23, what will happen?* Repeat for several other large numbers. Allow time for the students to complete the blackline master and share their answers.

[Properties]

Piggy Bank

Name _____

1. a. Draw 5 more coins. Then cross out 5 coins and complete the number sentence.

$\boxed{\text{\small ($1) ($1) ($1)}}$

($1) ($1) ($1)
($1) ($1) ($1)
$6 + \underline{\hspace{1cm}} - \underline{\hspace{1cm}} = 6$

b. Cross out 7 coins. Then complete the number sentence.

($1) ($1) ($1) ($1) ($1)   ($1) ($1) ($1) ($1) ($1)
($1) ($1) ($1) ($1) ($1)   ($1) ($1)

$10 + \underline{\hspace{0.3cm}7\hspace{0.3cm}} - \underline{\hspace{1cm}} = \underline{\hspace{1cm}}$

2. a. Draw jumps on the number line to show $23 + 4 - 4 = 23$.

b. Draw jumps to show $3 + 28 - 28 = 3$.

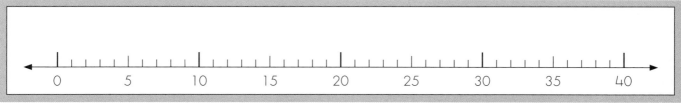

c. Draw jumps to show $15 + 7 - 7 = 15$.

3. Write numbers to make these true.

a. $12 + 5 - \underline{\hspace{1cm}} = 12$

b. $\underline{\hspace{1cm}} + 2 - 2 = 11$

c. $7 + \underline{\hspace{1cm}} - 8 = 7$

d. $9 - 6 + 6 = \underline{\hspace{1cm}}$

e. $13 + 9 - \underline{\hspace{1cm}} = 13$

f. $24 + \underline{\hspace{1cm}} - 10 = 24$

Turning Around

Representing addition turnarounds on a number line

AIM

Students will represent addition turnarounds on a number line and see that when two numbers are added in any order, the answer remains the same.

MATERIALS

- 2 sets of paper strips, each representing the numbers 1 to 10 (one blue set and one red set)

- Large number track for 1 to 20

- Blu-Tack

- 1 copy of the blackline master (opposite) for each student

- Calculator for each student

REFLECTION

Say, *If we add two numbers in any order the answer remains the same. Is this always true?* Call on volunteers to share their ideas.

1 Place the number track on the floor and have the students sit on either side. Show the blue paper strip for 6 and the red paper strip for 7 and ask, *How long is the blue strip?* Ask a volunteer to walk 6 spaces on the number track. Ask, *How long is the red strip? How long are both strips in total?* Have the student walk from 6 to 13 on the number track. Repeat for 7 + 2 and 8 + 7, with different students walking the number track.

2 Stick the blue paper strip for 3 horizontally on the board and then stick the red paper strip for 4 at the right, end to end. Ask, *How long is the blue strip plus the red strip?* On the board, below the strips, write **3 + 4 = 7**. Swap the order of the strips and ask, *How long is the red strip plus the blue strip?* Write **4 + 3 = 7** on the board below the first equation.

3 Draw a number line for 0 to 20 on the board. Ask, *Is 4 + 3 equal to 3 + 4? How can we use a number line to check?* Invite a volunteer to draw jumps above the number line to show 4 add 3. Ask another student to draw jumps on the underside of the number line to show 3 add 4, as shown below.

Say, *Both students finished on 7. Therefore, 4 + 3 is the same as or equal to 3 + 4.* Repeat for 8 + 5 and then 9 + 6.

4 Allow time for the students to complete the blackline master and share their answers.

[Properties]

Turning Around

1. Draw jumps to show how these number sentences are true.

   a. $12 + 6 = 6 + 12$

   b. $17 + 3 = 3 + 17$

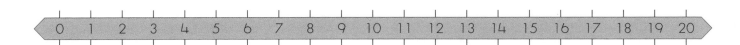

2. Write two different number sentences that show turnarounds.
   Draw jumps to show how your number sentences are true.

   a. _____ + _____ = _____ + _____

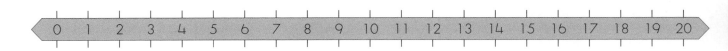

   b. _____ + _____ = _____ + _____

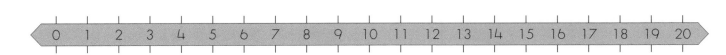

3. Use a calculator to check these number sentences. Shade true or false for each.

   a. $234 + 352 = 352 + 234$ ⬚ True ⬚ False

   b. $765 + 367 = 367 + 756$ ⬚ True ⬚ False

Properties

3

About Face

Subtracting numbers in different orders

AIM

Students will use number lines to represent subtraction involving two numbers and see that if the order is changed, the answer will be different.

MATERIALS

- 1 set of paper strips representing the numbers 1 to 10

- Large number track for −10 to 10

- 1 copy of the blackline master (opposite) for each student

- Calculator for each student

REFLECTION

Together discuss Question 3 on the blackline master. Ask, *When you subtract two numbers and change the order in which you subtract them, will the answer be different?*

1 Show the paper strip for 5 and the paper strip for 3. Ask, *What happens when we add the two strips and then change the order in which we add them?* (The answer remains the same.) *Will this work for subtraction?* Elicit several responses. Then show the strips for 5 and 3, and ask, *What is 5 take away 3?* Write **5 – 3 = 2** on the board. Show the two strips again and ask, *If we start with the strip for 3, can we take away 5? Why not?* (The strip is not long enough.) *If the temperature in this room was 8° and it fell by 5° overnight, what temperature would it be?* (3°) *How can we write this?* (8 – 5 = 3) *If the temperature was 5° and it fell by 8° overnight, what temperature would it be?* (Minus 3°.) *How can we write this?* (5 – 8 = –3) *Is 8 – 5 the same as 5 – 8?* (No.)

2 Draw a number line for −10 to 10 on the board. Invite a volunteer to show 8 – 5 on the number line by drawing jumps from 0 to 8 and then back 5. Ask another student to draw jumps to show 5 – 8, as shown below.

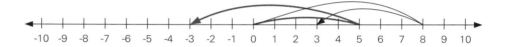

Ask, *Is 8 – 5 the same as 5 – 8?* (No. 8 – 5 = 3 and 5 – 8 = –3.) Repeat for 9 – 3 and then 7 – 4.

3 Ask the students to complete Question 1 on the blackline master. Together, complete Question 2a. Then ask, *Do you think 125 – 67 is the same as 67 – 125?* Allow the students to use a calculator to check. Ask the students to complete Question 3. Call on volunteers to share their answers.

[Properties]

About Face

Name _____

1. Draw jumps to show each of these.

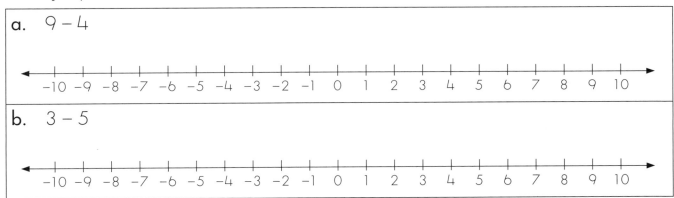

a. 9 – 4

```
←———+——+——+——+——+——+——+——+——+——+——+——+——+——+——+——+——+——+——+——+———→
   -10 -9 -8 -7 -6 -5 -4 -3 -2 -1  0  1  2  3  4  5  6  7  8  9  10
```

b. 3 – 5

```
←———+——+——+——+——+——+——+——+——+——+——+——+——+——+——+——+——+——+——+——+———→
   -10 -9 -8 -7 -6 -5 -4 -3 -2 -1  0  1  2  3  4  5  6  7  8  9  10
```

2. Draw jumps to show each of these. Write the answers. Shade true or false.

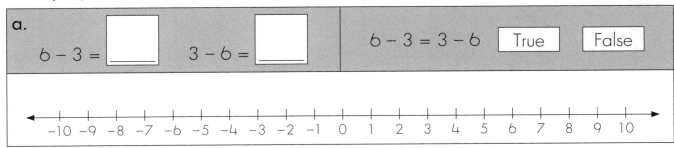

a.

6 – 3 = ☐   3 – 6 = ☐   6 – 3 = 3 – 6 [True] [False]

```
←———+——+——+——+——+——+——+——+——+——+——+——+——+——+——+——+——+——+——+——+———→
   -10 -9 -8 -7 -6 -5 -4 -3 -2 -1  0  1  2  3  4  5  6  7  8  9  10
```

b.

19 – 3 = ☐   3 – 19 = ☐   19 – 3 = 3 – 19 [True] [False]

```
←—|||||||||||||||||||||||||||||||||||||||||||—→
 -20        -10         0         10        20
```

3. Use a calculator to check these equations. Shade true or false.

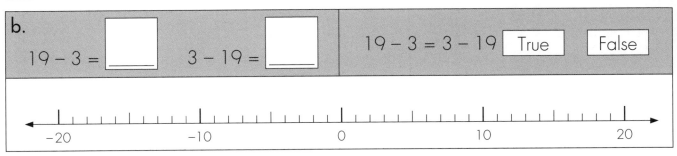

234 – 352 = 352 – 234   [True] [False]

765 – 367 = 367 – 765   [True] [False]

251 – 101 = 101 – 251   [True] [False]

Properties

4

Making Tracks

Adding numbers in different orders

AIM

Students will see that when any three numbers are added in any order the answer is the same.

MATERIALS

- Large number track for 0 to 20
- 1 copy of the blackline master (opposite) for each student
- Calculator for each student

REFLECTION

Challenge the students to think of real-life situations where, if they add the last two numbers first, they will get the same answer. Examples could include adding prices at the grocery store.

1 Place the number track on the floor. On the board, write **5 + 7 + 4**. Call on volunteers to share their answers to this addition problem. Ask them to share how they added the numbers. For example, did they add 5 and 7 and then 4? Ask, *Does it matter if we add the first 2 numbers and then the last number, or add the last 2 numbers and then the first number? Will we get the same answer?* Call on a volunteer to walk 5 + 7 and then 4 more, and another to walk 7 + 4 and then 5 more on the number track.

2 On the board, write **5 + 7 + 4 = 5 + 7 + 4** and ask, *Why did we link "5 add 7" and "7 add 4"?* (To record what numbers we added first.)

3 Ask the students to complete Question 1 on the blackline master. Call on volunteers to share their answers. Write their ideas on the board. Say, *Let's see if this works for large numbers.*

4 Ask the students to select any 3 numbers that are less than 100. Write the numbers on the board, for example **12 + 34 + 26 = 12 + 34 + 26**. Ask, *If we add the last 2 numbers and then the first number, will we have the same answer as if we added the first 2 numbers and then the last number?* Allow the students to check on a calculator. Repeat for 2 other sets of numbers.

5 Have the students complete the blackline master. Call on volunteers to share their answers.

Making Tracks

Name _____

1. Draw jumps to show these equations. Add the linked numbers first.

    a. $4 + 8 + 5 = 4 + 8 + 5$

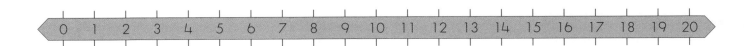

    b. $3 + 8 + 6 = 3 + 8 + 6$

    c. What did you notice above? _____

    _____

Properties

5

2. Use a calculator to check these equations, then shade true or false.

a.	$99 + 44 + 33 = 99 + 44 + 33$	True	False
b.	$56 + 71 + 12 = 56 + 71 + 12$	True	False
c.	$38 + 27 + 14 = 38 + 27 + 14$	True	False
d.	$23 + 68 + 31 = 23 + 68 + 31$	True	False
e.	$234 + 352 + 128 = 234 + 352 + 128$	True	False

More Then Less

Increasing and decreasing parts to leave addition equations unchanged

AIM

Students will explore the part–part–total pattern for addition equations; that is, if one part of an addition equation is increased then, in order to obtain the same answer, the other part must be decreased by the same amount.

MATERIALS

- 1 copy of the blackline master (opposite) for each student

REFLECTION

Discuss the idea that, *In addition, to keep the same answer, if we increase one part by an amount we must decrease the other part by the same amount.*

1 Say, *There are 6 parrots and 5 pigeons in a tree. How many birds are there in total? How do we write this?* (6 + 5 = 11) *If 3 more parrots fly to the tree, how many pigeons need to fly away so that there will still be the same number of birds in the tree?* (3) *How can we write this?* (6 + 3 + 5 – 3 = 11) *If we increase 6 by 4, how much must we decrease 5 by to keep the same amount?* (4) Call on volunteers to share their ideas.

2 Write **9 + 5 = 14** on the board and draw a number strip as shown below to represent the equation.

Ask, *If we increase 9 by 2, how do we change 5 so that the answer remains 14? Do we increase or decrease it? By how much?* Adjust the model, as shown, and call on a volunteer to write the new equation **11 + 3 = 14** on the board.

Repeat for increasing 9 by 3, 9 by 4, 5 by 6, and 5 by 2.

3 Read the blackline master with the class. Allow time for the students to complete the questions. Call on volunteers to share their answers.

More Then Less

Name _____

1.
$$3 + 12 = 15$$

a. Increase 3 by 4. Change 12 so that the answer is still 15. Shade this strip to show your answer. Write the new equation.

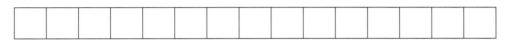

_____ + _____ = 15

b. Increase 3 by 8. Change 12 so that the answer is still 15. Shade this strip. Write the new equation.

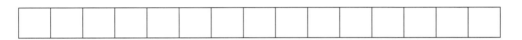

_____ + _____ = 15

2.
$$6 + 7 = 13$$

a. Increase 7 by 3. Change 6 so that the answer is still 13. Shade this strip. Write the new equation.

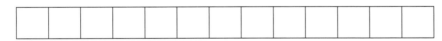

_____ + _____ = 13

b. Increase 7 by 4. Change 6 so that the answer is still 13. Shade this strip. Write the new equation.

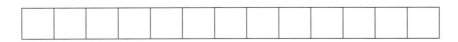

_____ + _____ = 13

Question Time

Exploring scatter plots to establish that the greater the distance from the horizontal axis, the greater the value of that point

AIM

Students will translate pictorial relationships presented in words on a scatter plot. They will also establish that the greater the distance from the horizontal axis, the greater the value of that point.

MATERIALS

- 1 copy of the blackline master (opposite) for each student

REFLECTION

Refer to the blackline master and ask, *In Question 1, how did you know who had the greatest number of sisters?* (The person at the greatest distance from the horizontal axis.) *How did you know who had the fewest number of sisters?* (The person at the nearest point to the horizontal axis.) Discuss the idea that the greater the distance from the horizontal axis, the greater the value of that point.

1 Draw the scatter plot shown below, on the board.

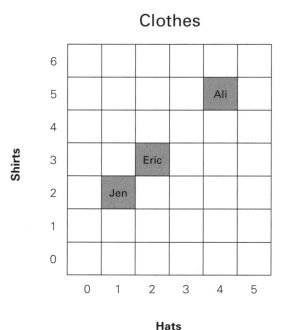

Ask, *What do the shaded parts of the graph represent?* (The number of hats and shirts each person has.) *What do the numbers across the bottom of the graph show?* (The number of hats.) *What do the numbers along the side of the graph show?* (The number of shirts.) *How many shirts does Jen have? How many hats does Ali have? Who has the greatest number of shirts? Who has the fewest number of hats? How many shirts are there in total? How many hats are there in total?* Make sure the students justify their answers.

2 Allow time for the students to complete the blackline master and share their answers.

Question Time

Name _____

---

1. Look at the graph.

   Families

   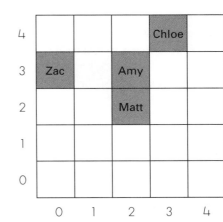

   a. Who has the most sisters? _____

   b. Who has no brothers? _____

   c. Who has the same number of sisters as brothers? _____

   d. Who has the largest family? _____

   e. Amy and Matt have _____ brothers in total.

---

2. Look at the graph.

   School Drawing

   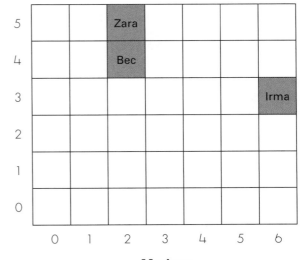

   a. Who has the most pencils? _____

   b. Who has the most markers? _____

   c. Who has 3 more markers than pencils? _____

   d. How many pencils and markers does Bec have in total? _____

   e. Zara and Irma have _____ markers in total.

How Many?

Exploring scatter plots to establish that the greater the distance
from the vertical axis, the greater the value of that point

AIM

Students will translate pictorial
relationships presented in words on
a scatter plot. They will also establish
that the greater the distance from the
vertical axis, the greater the value of
that point.

MATERIALS

• 1 copy of the blackline master
  (opposite) for each student

REFLECTION

Refer to the blackline master and
ask, *In Question 2, How did you
know who had the greatest number
of bikes?* (The person at the greatest
distance from the vertical axis.) *How
did you know who had the fewest
number of bikes?* (The person at the
nearest point to the vertical axis.)
Discuss the idea that the greater the
distance from the vertical axis, the
greater the value of that point.

1   Draw the scatter plot shown below, on the board.

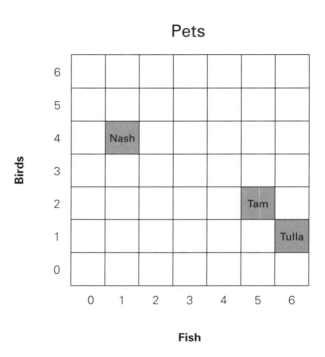

2   Ask questions such as, *What do the shaded parts of the graph
represent?* (The number of birds and fish each person has.) *What do
the numbers across the bottom of the graph show?* (The number of fish.)
What do the numbers along the side of the graph show? (The number
of birds.) *How many birds does Nash have? How many birds does Tam
have? Who has the greatest number of birds? Who has the fewest number
of birds? How many birds are there in total? Who has the fewest number
of fish? How many fish are there in total?* Call on several volunteers to
share and justify their answers.

3   Allow time for the students to complete the blackline master and share
their answers.

How Many?

Name _____

1. Look at the graph.

   a. Who has the greatest number of plants at home? _____

   b. Who has the greatest number of front steps at home? _____

   c. How many front steps are there in total? _____

   d. Who has more front steps than plants at home? _____

   e. Who has the same number of plants as front steps? _____

At Home

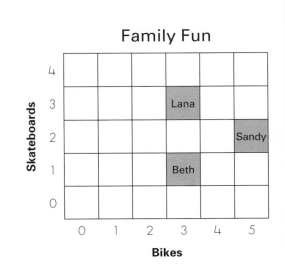

2. Look at the graph.

   a. Whose family has the greatest number of bikes? _____

   b. Whose family has the greatest number of skateboards? _____

   c. Whose family has the same number of skateboards as bikes? _____

   d. What is the total number of skateboards and bikes in Beth's family? _____

Family Fun

Representations

2

Seeing Pictures

Using drawings, number sentences, and tables to solve problems

AIM

Students will represent and solve simple real-world problems using drawings, number sentences, and tables. They will also begin to translate across different representations.

MATERIALS

- 1 copy of the blackline master (opposite) for each student

REFLECTION

Discuss the different ways to figure out problems (diagrams, tables, and numbers), and how each is related. Call on volunteers to share their ideas about which ways they prefer or which ways are more systematic.

1 Discuss how many wheels there are on bicycles and tricycles. Say, *Imagine there are some bicycles and tricycles in a shed. Each bicycle has 2 wheels, and each tricycle has 3 wheels. There are 22 wheels in total. How can we figure out how many bicycles and tricycles there are?* Discuss the following methods:

- On the board, draw a diagram of bicycles and then tricycles, and continually add to it until the number of wheels equals 22. Draw a similar diagram with the tricycles first, then the bicycles.

- Write addition number sentences to match the story, for example, $2 + 2 + 2 + 2 + 2 + 3 + 3 + 3 + 3$ or $3 + 3 + 3 + 3 + 3 + 3 + 2 + 2$.

- Draw tables, as shown below, and then select the numbers that add to 22.

Bicycles	Number of wheels
1	2
2	4
3	6
4	8
5	10
6	12

Tricycles	Number of wheels
1	3
2	6
3	9
4	12
5	15
6	18

Discuss how each representation is related.

2 Read and discuss each question on the blackline master. Allow time for the students to complete the questions. Call on volunteers to share which method of solving the problem they found to be the easiest.

Seeing Pictures

Name _____

There are some chairs and stools in the kitchen. There are 20 legs in total. How many chairs and stools are there?

1. Draw a picture to show the answer.

2. Complete these tables.

Number of chairs	Number of legs
1	
2	
3	
4	
5	

Number of stools	Number of legs
1	
2	
3	
4	
5	

3. Write an addition number sentence to show the answer.
    Use your picture or the tables to help.

_____

Representations

3

Growing Older

Using tables to describe change in real-world problems

AIM

Students will enter data in a table and use the table to describe real-world problems.

MATERIALS

- 1 copy of the blackline master (opposite) for each student

REFLECTION

Relate these problems to function machines. In each instance identify the addition change. Challenge the students to draw and complete tables about their own family. Have them write a series of questions relating to their tables. They can then ask another student to answer the questions.

1 Read Question 1 on the blackline master with the class, then draw the table on the board. Ask, *How many people are in the family?* For each person in the family ask, *How old is (Collette) now?* (12) *How old will she be in 5 years?* (17) Record the answers in the table. Ask the students to complete Questions 1 and 2. Allow time for them to share their answers.

2 Draw the following table on the board.

	Now	In ☐ years
Rikko	29	38
Marta	32	41
Lana	7	16

Say, *This is another family. Who is the father? Who is the mother?* Point to the unknown and ask, *What number should we write here? How do you know? Let's write a story about this family.* Allow the students to participate in writing the story. For example, *This is a family of three people: Rikko (the father), Marta (the mother), and Lara (the child). They are aged 29, 32, and 7 years. In 9 years, they will be 38, 41, and 16 years.*

3 Ask the students to complete the blackline master. Allow time for them to share their answers.

Growing Older

1. How old will this family be in 5 years?

   Figure out their ages and complete the table. Then answer the questions.

   How many people in the family? _____

   Who is the father? _____

   Who is the youngest child? _____

	Age now	Age in 5 years
Collette	12	
Max	9	
Emil	6	
Blanche	38	
Curtis	44	

2. How old will this family be in 7 years?

   Figure out their ages and complete the table. Then answer the questions.

   Who is the mother? _____

   Who is the oldest child? _____

	Age now	Age in 7 years
Jarvis	23	
Arabella	11	
Barrett	47	
Lola	53	

3. a. Complete the table.

   b. Write a story about this family.

   _____

   _____

   _____

   _____

   _____

   _____

	Age now	Age in 5 years
Nita	9	
Karl	6	
Sonja	38	

Representations

4

Mixed Sandwiches

Making ordered lists and recording in tables

AIM

Students will use three variables to systematically write an ordered list, and they will record this list in table format.

MATERIALS

- 1 copy of the blackline master (opposite) for each student
- Counters for each student

REFLECTION

Challenge the students to systematically find all the possible ways of placing 5 apples in 3 bowls. Allow them to use counters to help.

1 Read Question 1 on the blackline master with the class. On the board, draw a table with 12 rows and the following column headings: "Bread", "Filling", and "Extra". Ask, *If we use rye bread, what filling can we have? If we use salad filling, what extra can we have?* (Rye bread, salad, mustard; and rye bread, salad, mayonnaise.) Record the options in the table on the board and direct the students to record them in the table on the blackline master.

Ask, *If we use cheese filling, what extra can we have?* (Rye bread, cheese, mustard; and rye bread, cheese, mayonnaise.) Record these in the table. Allow time for the students to complete Question 1c and share their answers with the class.

2 Direct the students to complete Question 1d. Invite volunteers to share their answers with the class then ask, *How do you know if you have found all the ways? How did you record your answers systematically?* Elicit several responses.

[Representations]

Mixed Sandwiches

Name _____

1. How many different sandwiches can we make?

a. Choose rye bread with salad filling. Record the different sandwiches in the table.

b. Choose rye bread with cheese filling. Record the different sandwiches in the table.

c. Choose rye bread with egg filling. Record the different sandwiches in the table.

d. Complete the table using white bread with different fillings and extras.

Bread	Filling	Extra

Representations

5

ANSWERS

Equivalence and Equations 1 — Page 7

Building Balances
Name _____

1. Draw counters to balance the scales and complete the matching sentence or number sentence for each.

Model	Language	Number sentence
a.	Four add four equals six add two.	$4 + 4 = 6 + 2$
b.	Eight add three equals six add five.	$8 + 3 = 6 + 5$

2. Write numbers to balance the scales then complete the missing parts.

Model	Language	Number sentence
a. $5 + 6$ $8 + 3$	Five add six equals eight add three.	$5 + 6 = 8 + 3$
b. $2 + 5$ $6 + 1$	Two add five equals six add one.	$2 + 5 = 6 + 1$
c. $1 + 8$ $4 + 5$	One add eight equals four add five.	$1 + 8 = 4 + 5$

Equivalence and Equations 2 — Page 9

Feathered Friends
Name _____

Write numbers to show what you see in each tree picture.

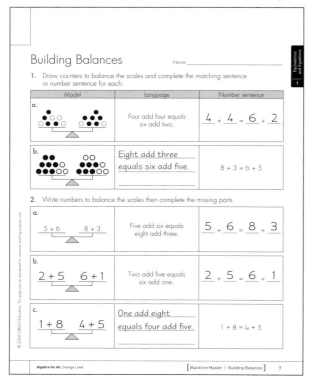
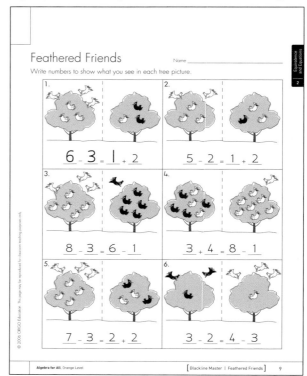

1. $6 - 3 = 1 + 2$
2. $5 - 2 = 1 + 2$
3. $8 - 3 = 6 - 1$
4. $3 + 4 = 8 - 1$
5. $7 - 3 = 2 + 2$
6. $3 - 2 = 4 - 3$

Equivalence and Equations 3 — Page 11

Same Sides
Name _____

Write the missing parts.

Model	Language	Number Sentence
1. $12 + 9$ $10 + 11$	12 add 9 equals 10 add 11.	$12 + 9 = 10 + 11$
2. 33 $29 + 4$	33 equals 29 add 4.	$33 = 29 + 4$
3. $34 - 5$ $27 + 2$	34 subtract 5 equals 27 add 2.	$34 - 5 = 27 + 2$
4. $67 - 9$ $56 + 2$	67 subtract 9 equals 56 add 2.	$67 - 9 = 56 + 2$
5. $34 + 15$ $55 - 6$	34 add 15 equals 55 subtract 6.	$34 + 15 = 55 - 6$

Equivalence and Equations 4 — Page 13

Balancing Act
Name _____

1. Write the missing number to make each scale balance.

a. $5 + 3 + 2$ $6 + 2$ | $5 + 3 + 2$ $6 + 2 + \boxed{2}$

b. $6 + 11 + 7$ $20 - 3$ | $6 + 11 + 7$ $20 - 3 + \boxed{7}$

c. $5 + 2$ $13 - 6 + 10$ | $5 + 2 + \boxed{10}$ $13 - 6 + 10$

2. Draw the missing jump and write the missing number to make the 2 sums equal.

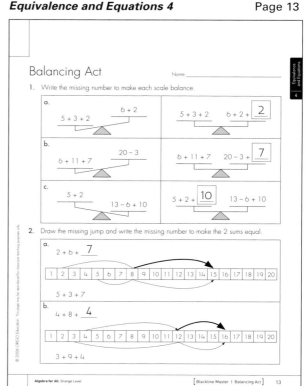

a. $2 + 6 + \underline{7}$

$5 + 3 + 7$

b. $4 + 8 + \underline{4}$

$3 + 9 + 4$

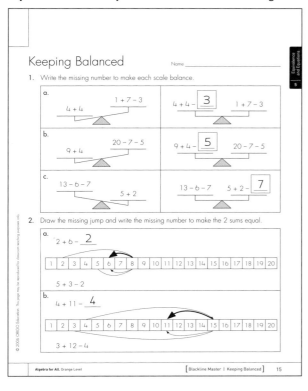

Keeping Balanced
Name _____

1. Write the missing number to make each scale balance.

a.
$4 + 4$ $1 + 7 - 3$
$4 + 4 - \boxed{3}$ $1 + 7 - 3$

b.
$9 + 4$ $20 - 7 - 5$
$9 + 4 - \boxed{5}$ $20 - 7 - 5$

c.
$13 - 6 - 7$ $5 + 2$
$13 - 6 - 7$ $5 + 2 - \boxed{7}$

2. Draw the missing jump and write the missing number to make the 2 sums equal.

a.
$2 + 6 - \underline{2}$
| 1 | 2 | 3 | 4 | 5 | 6 | 7 | 8 | 9 | 10 | 11 | 12 | 13 | 14 | 15 | 16 | 17 | 18 | 19 | 20 |
$5 + 3 - 2$

b.
$4 + 11 - \underline{4}$
| 1 | 2 | 3 | 4 | 5 | 6 | 7 | 8 | 9 | 10 | 11 | 12 | 13 | 14 | 15 | 16 | 17 | 18 | 19 | 20 |
$3 + 12 - 4$

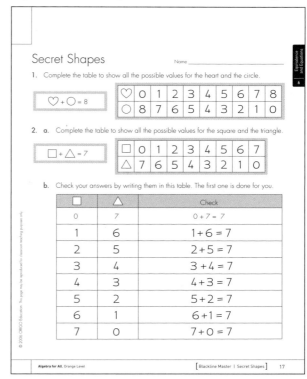

Secret Shapes
Name _____

1. Complete the table to show all the possible values for the heart and the circle.

$\heartsuit + \bigcirc = 8$

♡	0	1	2	3	4	5	6	7	8
○	8	7	6	5	4	3	2	1	0

2. a. Complete the table to show all the possible values for the square and the triangle.

$\square + \triangle = 7$

□	0	1	2	3	4	5	6	7
△	7	6	5	4	3	2	1	0

b. Check your answers by writing them in this table. The first one is done for you.

□	△	Check
0	7	$0 + 7 = 7$
1	6	$1 + 6 = 7$
2	5	$2 + 5 = 7$
3	4	$3 + 4 = 7$
4	3	$4 + 3 = 7$
5	2	$5 + 2 = 7$
6	1	$6 + 1 = 7$
7	0	$7 + 0 = 7$

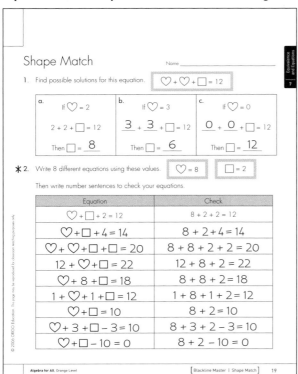

Shape Match
Name _____

1. Find possible solutions for this equation. $\heartsuit + \heartsuit + \square = 12$

a.
If $\heartsuit = 2$
$2 + 2 + \square = 12$
Then $\square = \underline{8}$

b.
If $\heartsuit = 3$
$\underline{3} + \underline{3} + \square = 12$
Then $\square = \underline{6}$

c.
If $\heartsuit = 0$
$\underline{0} + \underline{0} + \square = 12$
Then $\square = \underline{12}$

✶2. Write 8 different equations using these values. $\heartsuit = 8$ $\square = 2$

Then write number sentences to check your equations.

Equation	Check
$\heartsuit + \square + 2 = 12$	$8 + 2 + 2 = 12$
$\heartsuit + \square + 4 = 14$	$8 + 2 + 4 = 14$
$\heartsuit + \heartsuit + \square + \square = 20$	$8 + 8 + 2 + 2 = 20$
$12 + \heartsuit + \square = 22$	$12 + 8 + 2 = 22$
$\heartsuit + 8 + \square = 18$	$8 + 8 + 2 = 18$
$1 + \heartsuit + 1 + \square = 12$	$1 + 8 + 1 + 2 = 12$
$\heartsuit + \square = 10$	$8 + 2 = 10$
$\heartsuit + 3 + \square - 3 = 10$	$8 + 3 + 2 - 3 = 10$
$\heartsuit + \square - 10 = 0$	$8 + 2 - 10 = 0$

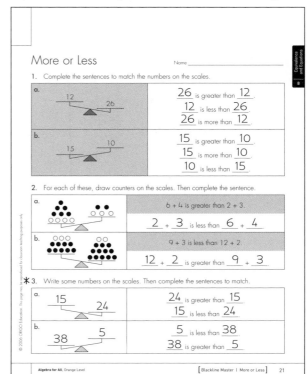

More or Less
Name _____

1. Complete the sentences to match the numbers on the scales.

a.
12   26
26 is greater than 12.
12 is less than 26.
26 is more than 12.

b.
15   10
15 is greater than 10.
15 is more than 10.
10 is less than 15.

2. For each of these, draw counters on the scales. Then complete the sentence.

a.
$6 + 4$ is greater than $2 + 3$.
$\underline{2} + \underline{3}$ is less than $\underline{6} + \underline{4}$.

b.
$9 + 3$ is less than $12 + 2$.
$\underline{12} + \underline{2}$ is greater than $\underline{9} + \underline{3}$.

✶3. Write some numbers on the scales. Then complete the sentences to match.

a.
15   24
24 is greater than 15.
15 is less than 24.

b.
38   5
5 is less than 38.
38 is greater than 5.

✶ Answers will vary. This is one example.

ANSWERS

Patterns and Functions 1   Page 23

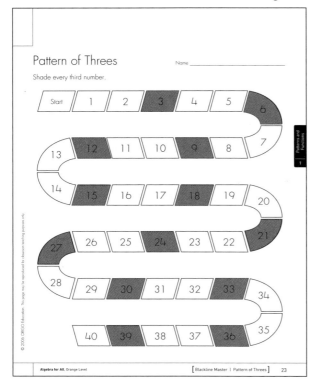

Pattern of Threes

Shade every third number.

Patterns and Functions 2   Page 25

Repeating Repeating

1. a. Write the number of shapes in each repeat. The first one is done for you.

   4   4   4   4   4

   b. Draw the 12th shape. ◯   c. Draw the 20th shape. ◯

   d. Draw the 14th shape. ☐   e. Draw the 17th shape. ☐

   f. Imagine the pattern continues. Draw the 45th shape. ☐

   ✱ Write how you figured it out. The 4th shape in each repeat is a circle. If I count by 4s, the 44th shape is a circle so the 45th shape is a square.

2. a. Write the number of shapes in each repeat.

   5   5   5

   b. Draw the 9th shape. ◯   c. Draw the 5th shape. △

   d. Draw the 15th shape. △   e. Draw the 11th shape. △

   f. Imagine the pattern continues. Draw the 31st shape. △

   ✱ Write how you figured it out. The 5th shape in each repeat is a triangle. If I count by 5s, the 30th shape is a triangle. 31 is the 1st shape in the next repeat, so it is a triangle.

Patterns and Functions 3   Page 27

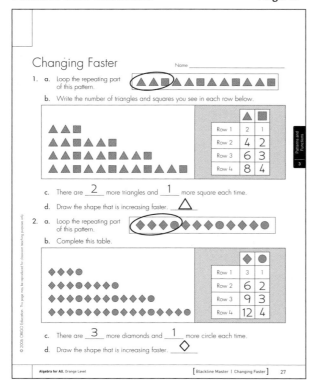

Changing Faster

1. a. Loop the repeating part of this pattern.

   b. Write the number of triangles and squares you see in each row below.

   c. There are 2 more triangles and 1 more square each time.

   d. Draw the shape that is increasing faster. △

2. a. Loop the repeating part of this pattern.

   b. Complete this table.

   c. There are 3 more diamonds and 1 more circle each time.

   d. Draw the shape that is increasing faster. ◇

Patterns and Functions 4   Page 29

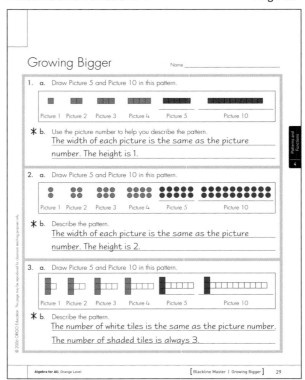

Growing Bigger

1. a. Draw Picture 5 and Picture 10 in this pattern.

   Picture 1   Picture 2   Picture 3   Picture 4   Picture 5   Picture 10

   ✱ b. Use the picture number to help you describe the pattern.
   The width of each picture is the same as the picture number. The height is 1.

2. a. Draw Picture 5 and Picture 10 in this pattern.

   Picture 1   Picture 2   Picture 3   Picture 4   Picture 5   Picture 10

   ✱ b. Describe the pattern.
   The width of each picture is the same as the picture number. The height is 2.

3. a. Draw Picture 5 and Picture 10 in this pattern.

   Picture 1   Picture 2   Picture 3   Picture 4   Picture 5   Picture 10

   ✱ b. Describe the pattern.
   The number of white tiles is the same as the picture number. The number of shaded tiles is always 3.

✱ Answers will vary. This is one example.

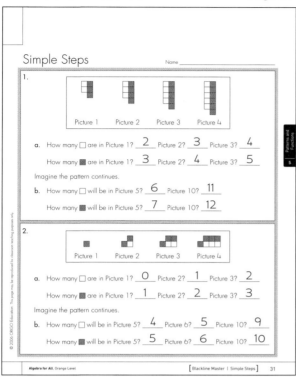

Simple Steps   Name _____

1.

Picture 1   Picture 2   Picture 3   Picture 4

a. How many ☐ are in Picture 1? __2__ Picture 2? __3__ Picture 3? __4__

How many ■ are in Picture 1? __3__ Picture 2? __4__ Picture 3? __5__

Imagine the pattern continues.

b. How many ☐ will be in Picture 5? __6__ Picture 10? __11__

How many ■ will be in Picture 5? __7__ Picture 10? __12__

2.

Picture 1   Picture 2   Picture 3   Picture 4

a. How many ☐ are in Picture 1? __0__ Picture 2? __1__ Picture 3? __2__

How many ■ are in Picture 1? __1__ Picture 2? __2__ Picture 3? __3__

Imagine the pattern continues.

b. How many ☐ will be in Picture 5? __4__ Picture 6? __5__ Picture 10? __9__

How many ■ will be in Picture 5? __5__ Picture 6? __6__ Picture 10? __10__

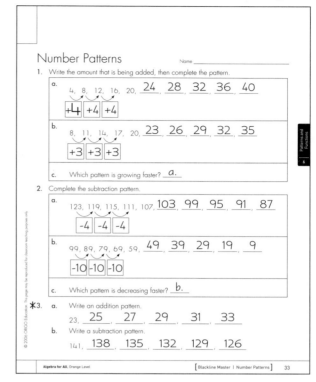

Number Patterns   Name _____

1. Write the amount that is being added, then complete the pattern.

a. 4, 8, 12, 16, 20, __24__ __28__ __32__ __36__ __40__

+4 +4 +4

b. 8, 11, 14, 17, 20, __23__ __26__ __29__ __32__ __35__

+3 +3 +3

c. Which pattern is growing faster? __a.__

2. Complete the subtraction pattern.

a. 123, 119, 115, 111, 107, __103__ __99__ __95__ __91__ __87__

-4 -4 -4

b. 99, 89, 79, 69, 59, __49__ __39__ __29__ __19__ __9__

-10 -10 -10

c. Which pattern is decreasing faster? __b.__

✱3. a. Write an addition pattern.

23, __25__ __27__ __29__ __31__ __33__

b. Write a subtraction pattern.

141, __138__ __135__ __132__ __129__ __126__

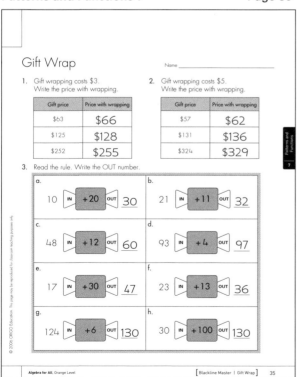

Gift Wrap   Name _____

1. Gift wrapping costs $3. Write the price with wrapping.

Gift price	Price with wrapping
$63	$66
$125	$128
$252	$255

2. Gift wrapping costs $5. Write the price with wrapping.

Gift price	Price with wrapping
$57	$62
$131	$136
$324	$329

3. Read the rule. Write the OUT number.

a. 10 IN +20 OUT 30

b. 21 IN +11 OUT 32

c. 48 IN +12 OUT 60

d. 93 IN +4 OUT 97

e. 17 IN +30 OUT 47

f. 23 IN +13 OUT 36

g. 124 IN +6 OUT 130

h. 30 IN +100 OUT 130

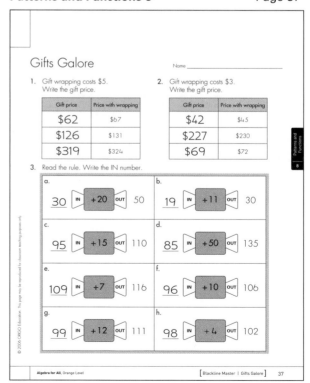

Gifts Galore   Name _____

1. Gift wrapping costs $5. Write the gift price.

Gift price	Price with wrapping
$62	$67
$126	$131
$319	$324

2. Gift wrapping costs $3. Write the gift price.

Gift price	Price with wrapping
$42	$45
$227	$230
$69	$72

3. Read the rule. Write the IN number.

a. 30 IN +20 OUT 50

b. 19 IN +11 OUT 30

c. 95 IN +15 OUT 110

d. 85 IN +50 OUT 135

e. 109 IN +7 OUT 116

f. 96 IN +10 OUT 106

g. 99 IN +12 OUT 111

h. 98 IN +4 OUT 102

✱ Answers will vary. This is one example.

ANSWERS

Patterns and Functions 9   Page 39

More Gifts   Name _____

1. For each of these, figure out the cost of gift wrapping.

a. Cost of wrapping **$20**

Gift price	Price with wrapping
$47	$67
$111	$131
$304	$324

b. Cost of wrapping **$7**

Gift price	Price with wrapping
$25	$32
$30	$37
$76	$83

2. Write the rules.

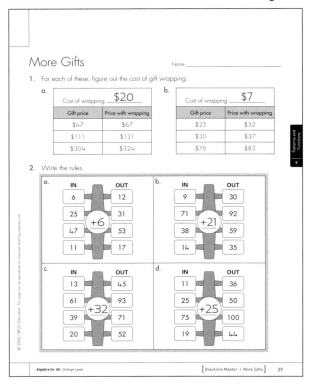

Algebra for All, Orange Level   [Blackline Master | More Gifts]   39

Patterns and Functions 10   Page 41

Super Savings   Name _____

1. A restaurant gives out $4 coupons. Write the new price of each meal.

Meal price	New price
$23	**$19**
$57	**$53**
$41	**$37**

2. A restaurant gives out $7 coupons. Write the new price of each meal.

Meal price	New price
$25	**$18**
$39	**$32**
$56	**$49**

3. Read the rule. Write the OUT number.

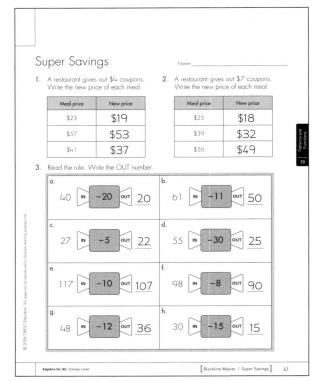

Algebra for All, Orange Level   [Blackline Master | Super Savings]   41

Patterns and Functions 11   Page 43

Meal Deals   Name _____

1. A coupon gives $5 off meals. Write the full price of each meal.

Full meal price	Price with coupon
$72	$67
$36	$31
$29	$24

2. A coupon gives $3 off meals. Write the full price of each meal.

Full meal price	Price with coupon
$48	$45
$33	$30
$75	$72

3. Read the rule. Write the IN number.

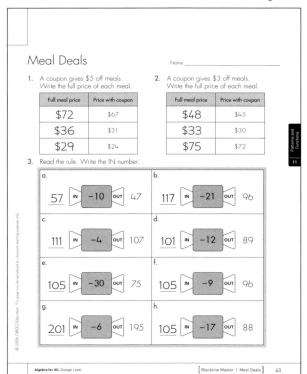

Algebra for All, Orange Level   [Blackline Master | Meal Deals]   43

Properties 1   Page 45

Zero Identity   Name _____

1. a. Cross out 17 counters. Write the matching number sentence.

17 – 17 = 0

b. Cross out $8. Write the matching number sentence.

8 – 8 = 0

c. Draw 10 balls. Cross out 10 balls. Write the matching number sentence.

17 – 17 = 0

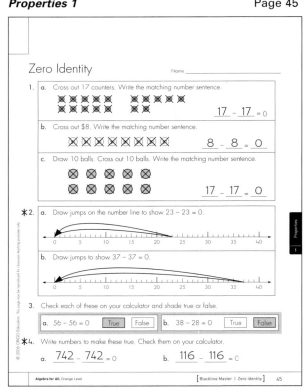

*2. a. Draw jumps on the number line to show 23 – 23 = 0.

b. Draw jumps to show 37 – 37 = 0.

3. Check each of these on your calculator and shade true or false.

a. 56 – 56 = 0   [**True**] [False]   b. 38 – 28 = 0   [True] [**False**]

*4. Write numbers to make these true. Check them on your calculator.

a. **742 – 742** = 0   b. **116 – 116** = 0

Algebra for All, Orange Level   [Blackline Master | Zero Identity]   45

---

* Answers will vary. This is one example.

Properties 2

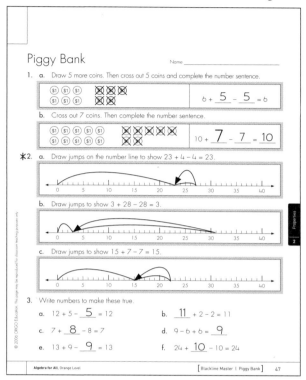

Piggy Bank
Name _____

1. a. Draw 5 more coins. Then cross out 5 coins and complete the number sentence.

$6 + \underline{5} - \underline{5} = 6$

  b. Cross out 7 coins. Then complete the number sentence.

$10 + \underline{7} - \underline{7} = 10$

✱2. a. Draw jumps on the number line to show $23 + 4 - 4 = 23$.

  b. Draw jumps to show $3 + 28 - 28 = 3$.

  c. Draw jumps to show $15 + 7 - 7 = 15$.

3. Write numbers to make these true.

  a. $12 + 5 - \underline{5} = 12$ b. $\underline{11} + 2 - 2 = 11$

  c. $7 + \underline{8} - 8 = 7$ d. $9 - 6 + 6 = \underline{9}$

  e. $13 + 9 - \underline{9} = 13$ f. $24 + \underline{10} - 10 = 24$

Properties 3

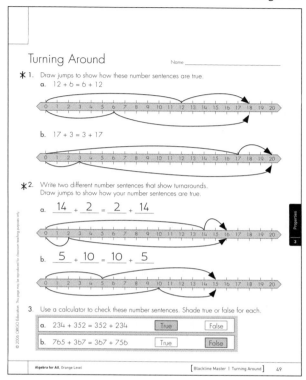

Turning Around
Name _____

✱1. Draw jumps to show how these number sentences are true.
  a. $12 + 6 = 6 + 12$

  b. $17 + 3 = 3 + 17$

✱2. Write two different number sentences that show turnarounds.
  Draw jumps to show how your number sentences are true.

  a. $\underline{14} + \underline{2} = \underline{2} + \underline{14}$

  b. $\underline{5} + \underline{10} = \underline{10} + \underline{5}$

3. Use a calculator to check these number sentences. Shade true or false for each.

  a. $234 + 352 = 352 + 234$ [True]   False

  b. $765 + 367 = 367 + 756$ True   [False]

Properties 4

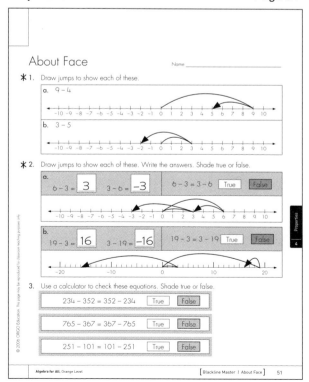

About Face
Name _____

✱1. Draw jumps to show each of these.
  a. $9 - 4$

  b. $3 - 5$

✱2. Draw jumps to show each of these. Write the answers. Shade true or false.

  a. $6 - 3 = \boxed{3}$ $3 - 6 = \boxed{-3}$ $6 - 3 = 3 - 6$ True [False]

  b. $19 - 3 = \boxed{16}$ $3 - 19 = \boxed{-16}$ $19 - 3 = 3 - 19$ True [False]

3. Use a calculator to check these equations. Shade true or false.

  $234 - 352 = 352 - 234$ True [False]

  $765 - 367 = 367 - 765$ True [False]

  $251 - 101 = 101 - 251$ True [False]

Properties 5

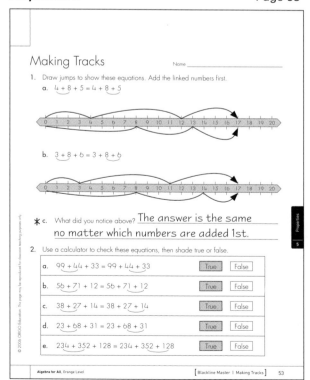

Making Tracks
Name _____

1. Draw jumps to show these equations. Add the linked numbers first.
  a. $4 + 8 + 5 = 4 + 8 + 5$

  b. $3 + 8 + 6 = 3 + 8 + 6$

✱c. What did you notice above? The answer is the same no matter which numbers are added 1st.

2. Use a calculator to check these equations, then shade true or false.

  a. $99 + 44 + 33 = 99 + 44 + 33$ [True]   False

  b. $56 + 71 + 12 = 56 + 71 + 12$ [True]   False

  c. $38 + 27 + 14 = 38 + 27 + 14$ [True]   False

  d. $23 + 68 + 31 = 23 + 68 + 31$ [True]   False

  e. $234 + 352 + 128 = 234 + 352 + 128$ [True]   False

✱ Answers will vary. This is one example.

ANSWERS

Properties 6
Page 55

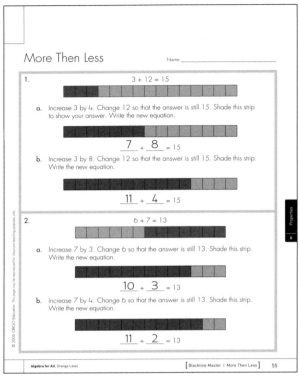

More Then Less
Name _____

1.
$$3 + 12 = 15$$

a. Increase 3 by 4. Change 12 so that the answer is still 15. Shade this strip to show your answer. Write the new equation.

$$\underline{7} + \underline{8} = 15$$

b. Increase 3 by 8. Change 12 so that the answer is still 15. Shade this strip. Write the new equation.

$$\underline{11} + \underline{4} = 15$$

2.
$$6 + 7 = 13$$

a. Increase 7 by 3. Change 6 so that the answer is still 13. Shade this strip. Write the new equation.

$$\underline{10} + \underline{3} = 13$$

b. Increase 7 by 4. Change 6 so that the answer is still 13. Shade this strip. Write the new equation.

$$\underline{11} + \underline{2} = 13$$

Representations 1
Page 57

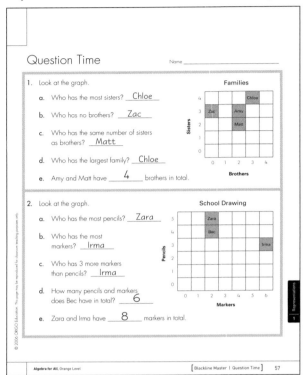

Question Time
Name _____

1. Look at the graph.

    Families

    a. Who has the most sisters? __Chloe__

    b. Who has no brothers? __Zac__

    c. Who has the same number of sisters as brothers? __Matt__

    d. Who has the largest family? __Chloe__

    e. Amy and Matt have __4__ brothers in total.

2. Look at the graph.

    School Drawing

    a. Who has the most pencils? __Zara__

    b. Who has the most markers? __Irma__

    c. Who has 3 more markers than pencils? __Irma__

    d. How many pencils and markers does Bec have in total? __6__

    e. Zara and Irma have __8__ markers in total.

Representations 2
Page 59

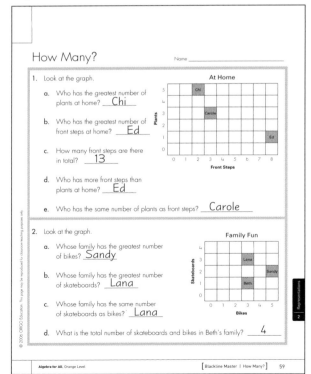

How Many?
Name _____

1. Look at the graph.

    At Home

    a. Who has the greatest number of plants at home? __Chi__

    b. Who has the greatest number of front steps at home? __Ed__

    c. How many front steps are there in total? __13__

    d. Who has more front steps than plants at home? __Ed__

    e. Who has the same number of plants as front steps? __Carole__

2. Look at the graph.

    Family Fun

    a. Whose family has the greatest number of bikes? __Sandy__

    b. Whose family has the greatest number of skateboards? __Lana__

    c. Whose family has the same number of skateboards as bikes? __Lana__

    d. What is the total number of skateboards and bikes in Beth's family? __4__

Representations 3
Page 61

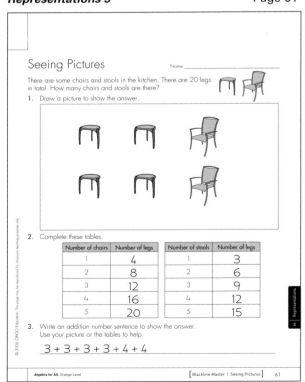

Seeing Pictures
Name _____

There are some chairs and stools in the kitchen. There are 20 legs in total. How many chairs and stools are there?

1. Draw a picture to show the answer.

2. Complete these tables.

Number of chairs	Number of legs
1	4
2	8
3	12
4	16
5	20

Number of stools	Number of legs
1	3
2	6
3	9
4	12
5	15

3. Write an addition number sentence to show the answer. Use your picture or the tables to help.

$$3 + 3 + 3 + 3 + 4 + 4$$

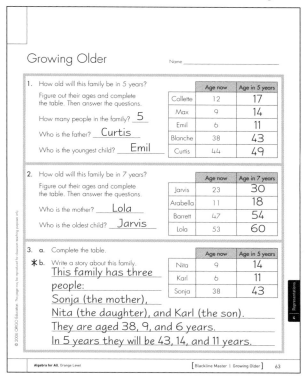

Growing Older

Name _____

1. How old will this family be in 5 years?

    Figure out their ages and complete the table. Then answer the questions.

    How many people in the family? __5__

    Who is the father? __Curtis__

    Who is the youngest child? __Emil__

	Age now	Age in 5 years
Collette	12	17
Max	9	14
Emil	6	11
Blanche	38	43
Curtis	44	49

2. How old will this family be in 7 years?

    Figure out their ages and complete the table. Then answer the questions.

    Who is the mother? __Lola__

    Who is the oldest child? __Jarvis__

	Age now	Age in 7 years
Jarvis	23	30
Arabella	11	18
Barrett	47	54
Lola	53	60

3. a. Complete the table.

    ✶ b. Write a story about this family.

    This family has three people:
    Sonja (the mother),
    Nita (the daughter), and Karl (the son).
    They are aged 38, 9, and 6 years.
    In 5 years they will be 43, 14, and 11 years.

	Age now	Age in 5 years
Nita	9	14
Karl	6	11
Sonja	38	43

Mixed Sandwiches

Name _____

1. How many different sandwiches can we make?

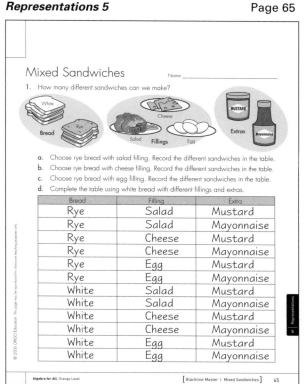

a. Choose rye bread with salad filling. Record the different sandwiches in the table.
b. Choose rye bread with cheese filling. Record the different sandwiches in the table.
c. Choose rye bread with egg filling. Record the different sandwiches in the table.
d. Complete the table using white bread with different fillings and extras.

Bread	Filling	Extra
Rye	Salad	Mustard
Rye	Salad	Mayonnaise
Rye	Cheese	Mustard
Rye	Cheese	Mayonnaise
Rye	Egg	Mustard
Rye	Egg	Mayonnaise
White	Salad	Mustard
White	Salad	Mayonnaise
White	Cheese	Mustard
White	Cheese	Mayonnaise
White	Egg	Mustard
White	Egg	Mayonnaise

✶ Answers will vary. This is one example.

Assessment Summary

Name _____

	Lesson	Page	A	B	C	D	Date
Equivalence and Equations	Building Balances	6					
	Feathered Friends	8					
	Same Sides	10					
	Balancing Act	12					
	Keeping Balanced	14					
	Secret Shapes	16					
	Shape Match	18					
	More or Less	20					
Patterns and Functions	Pattern of Threes	22					
	Repeating Repeating	24					
	Changing Faster	26					
	Growing Bigger	28					
	Simple Steps	30					
	Number Patterns	32					
	Gift Wrap	34					
	Gifts Galore	36					
	More Gifts	38					
	Super Savings	40					
	Meal Deals	42					
Properties	Zero Identity	44					
	Piggy Bank	46					
	Turning Around	48					
	About Face	50					
	Making Tracks	52					
	More Then Less	54					
Representations	Question Time	56					
	How Many?	58					
	Seeing Pictures	60					
	Growing Older	62					
	Mixed Sandwiches	64					

Algebra for All, Orange Level